Medical Marvels

Medical Marvels

The *100 Greatest* Advances in Medicine

Eugene W. Straus, MD,
& Alex Straus

Illustrations by Bette Korman

Prometheus Books

59 John Glenn Drive
Amherst, New York 14228-2197

Published 2006 by Prometheus Books

Inquiries should be addressed to
Prometheus Books
59 John Glenn Drive
Amherst, New York 14228–2197
VOICE: 716–691–0133, ext. 207
FAX: 716–564–2711
WWW.PROMETHEUSBOOKS.COM

10 09 08 07 06   5 4 3 2 1

Library of Congress Cataloging-in-Publication Data

Straus, Eugene.
    Medical marvels : the 100 greatest advances in medicine / by Eugene W. Straus and Alex Straus; illustrations by Bette Korman.
        p. cm.
    Includes bibliographical references and index.
    ISBN 1–59102–373–4 (hardcover : alk. paper)
    1. Medicine—History. 2. Medical innovations—History. I. Straus, Alex. II. Title.

R131.S83 2006
610—dc22   2005027624

Printed in the United States of America on acid-free paper

For Bette,

who is my rock, and my roll

In memory of

Bernard Straus, MD, Pearl Whitehorn Straus,
Mark Straus, MD, Robert Krasnow, MD,
Israel Miller, MD, Solomon Berson, MD, and Rachmeil Levine, MD

Contents

As Prometheus gave humankind fire, and with it wisdom,
so Apollo, the physician,
gave us healing,
and with it compassion.

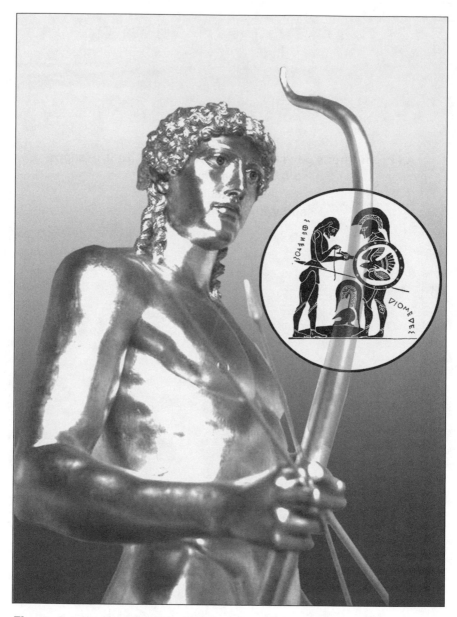

Fig. 1. Apollo the Physician and an inset of a Greek healer
bandaging an injured Homeric warrior.

Acknowledgments

For their support and advice we thank the following individuals: Ms. Allegra Straus; Mrs. Sylvie Straus-Figueroa, Esq; Mr. Enrique Figueroa; Dr. Harold Gainer; Mrs. Ruth Gainer; Ms. Vivian Gainer; Mr. Benjamin Gainer, Esq.; Dr. Jesse Gainer; Ms. Clara Luna Gainer; Dr. Norman Krasnow; Mr. Stephan Kass, Esq; Ms. Irene Kellog; Mrs. Radha Korman Huszti; Mr. Douglas Huszti; Mr. Louis Silverstein; Ms. Helen Abby Becker; Dr. Stanley R.Yancovitz; Dr. Allan Frances; Mr. Benjamin Yalow; and Dr. Rosalyn S. Yalow.

We thank Beth Stadelberger and Lisa Slovensky for technical assistence.

We thank our brilliant agent, Robin Straus, who is no relation, but we wish she were.

We thank our editor, Linda Greenspan Regan, for her unfailing judgment, taste, and tact, and Jeremy Sauer for his meticulous attention and thoughtful guidance.

This effort owes much to the students, fellows, colleagues, and patients with whom I have worked during the past forty years.

Preface

I swear by Apollo the physician . . .

—The Hippocratic Oath

For of the most High cometh healing.

—Ecclesiasticus 38:2

This book took shape in response to a question I was asked while giving a Grand Rounds presentation at a medical school. I was lecturing about an uncommon hormonal disorder when a senior medical student rose to say that she had a patient who might have this disease. "I have read your papers, *professor*, it is all very *interesting*, but my patient cannot *afford* the diagnostic tests or treatments. And with the weaponization of disease, and the commerce in human organs, and the racism and sexism right here in this school, which I assure you is better than most, and . . . Dr. Straus, tell me why I should not quit medicine. What is so great about medicine?"

"Think of all the good you can do," I said, aware of how unhelpful that answer was, and excusing my tragicomic irrelevance by recognizing that, after all, I only had a few seconds to come up with an answer to this deeply perplexing question. The question stuck with me like a forlorn song. It presented itself differently in a variety of situations, until I began formulating better answers in my mind. And then I began to believe that the issues that kept arising were especially relevant now, in these healthcare-challenged times. So, after forty years of practicing medicine, of teaching as a professor of medicine and serving as department chair in four medical schools, of doing research at the "cutting edge" in elite settings, and traveling the world learning, teaching, and consulting with the health ministries of countries like

17

China and India, I decided to buckle down and put my thoughts down on computer. At first it was in the spirit of Gustave Flaubert, " . . . simply for the pleasure of writing, for myself alone. . . . Apollo at least will be grateful to me. . . ."[1] But then, as the project grew, I enlisted the collaboration of my son for the intelligent layman's point of view, and it didn't hurt that he is an accomplished journalist.

Among "the better angels of our nature," those concerned with the relief of suffering, healing, and prevention of illness, deserve a special place in our hearts. At this time, when scientific and technological advances frequently are seen as having destructive potential, and the very notion of progress may be questioned, we examine human ingenuity's most hallowed ground. Advances in the art and science of healing are among the great achievements of human culture, and to describe the greatest among them is an awesome challenge.

This book is an introduction to some of the great ideas, people, and accomplishments that have influenced the development of healing. It is a celebration of human intellect, spirit, and soul. It is not a history per se of medicine—it is equally concerned with our current climate, and we hope it is a call to thought and action. If you want to understand where medicine is going, if you want a powerful and accurate lens through which to view society, then we hope this book is for you. Our thesis is that we all must play a role to ensure that medicine is applied for the benefit of life. From the reaction I've received from both the medical and the lay community alike to lectures and articles, I can see that there is intense interest in these stories of discovery and the issues they raise. Healthcare is among the most significant issues of our time, and it will be more so in the future. Sad but true, healthcare is now a battle ground, and it is hazardous to have opinions and take action without background knowledge. More dangerous still is to sit on the sidelines as the destiny of your people is fought over and squandered.

Here we will celebrate the greatest advances in medicine. Our fundamental criterion for inclusion among the greatest developments is that the idea, treatment, device, institution, or tradition has had a major effect upon the lives of people. For the most part, we use "great" to mean eminent or excellent, thus connoting a positive value. Almost all these great advances have resulted in unprecedented improvements in health or alleviation of suffering. In some circumstances, however, *great* is used to imply intensity, or extent of hope and excitement behind a development that, while causing sweeping change, may be of questionable value. In this context, the advance represents the way medicine moved into the future, but not necessarily improved. Of course, with the passage of time the relative impact of certain advances can change. For example, the human genome project, so full of potential, has yet to make its ultimate mark in terms of its benefit. For this reason there will, and should, be legitmate debate regarding this list. Therefore it is important to undersand the standards by which we made our selections. Each advance had to satisfy one or more of the following four criteria:

1. It changed the course of medicine.
2. It resulted in profound relief of suffering.
3. It caused a fundamental improvement in the understanding of health and disease.
4. It produced great improvement in prevention, diagnosis, or treatment of disease.

Medicine is a uniquely human activity, so it should not surprise us that many of the greatest advances are human developments, rather than technological innovations. And because of medicine's human element, it is full of paradoxes and hard choices, as well as meaningful contributions from non-physician-scientists. In fact, the scientific community understands that it always needs the thoughts and opinions of people like you. We hope that you find our listing and discussions to be interesting, informative, controversial, occasionally infuriating, and subject to improvement. Choosing the 100 greatest advances was difficult. Ranking them in order of importance, as we do here, was impossible, perhaps even foolhardy, but we attempted it only for the purposes of contemplation and discussion, beginning with the most significant advancement through the next 100 in importance.

Enjoy, discuss, and get involved!
Eugene Straus, MD
West Shokan, NY
March 2005

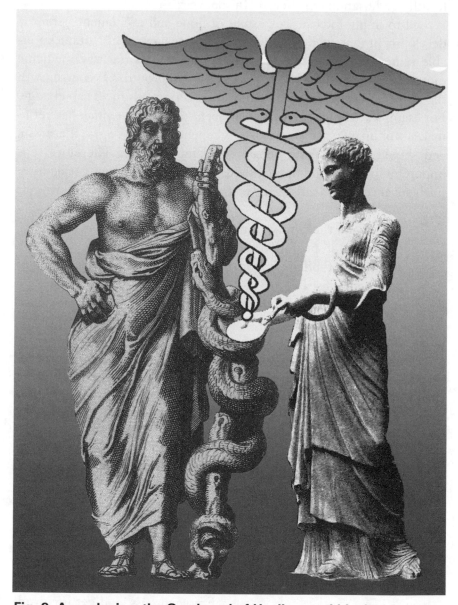

Fig. 2. Aesculapius, the Greek god of Healing, and his daughter, Hygiea, the Greek goddess of Health, with their snake-entwined staff that became the medical symbol known as Caduceus.

1.

From Abandonment
to the Idea of Healing

In primitive society, where uniformity of occupation is the
rule, and the distribution of the community into various
classes of workers has hardly begun, every man is more or
less his own magician; he practises charms and incanta-
tions for his own good and the injury of his enemies. But a
great step in advance has been taken when a special class
of magicians has been instituted; when, in other words, a
number of men have been set apart for the express purpose of benefiting the whole community by their skill,
whether that skill be directed to the healing of diseases,
the forecasting of the future, the regulation of the weather,
or any other object of general utility.

—Sir James George Frazer (1854–1941), *The Golden Bough*, 1922

ttitudes toward illness evolve and are, at all times and
places, inconstant. There have even been times in
human history when most of humanity refused to confront illness at all.

Anthropologists, who study the artifacts of prehistoric hunting
and gathering groups, find evidence of individual illness, but little sign
of devastating epidemics. The reason for this is believed to be a matter

21

of practicality. A lame or sickened individual would certainly pose a serious threat to a band or nomadic hunting-gathering group on the move, and the earliest responses to illness may well have been to leave the sufferer behind. After all, we see evidence of that in the treatment of lepers, ritually expunged from societies until recent times, extinguished while still alive. Abandonment is still an exercised option in the approach to the ill and the elderly.

Over centuries various notions regarding the origins and significance of illness have evolved—from those born of the natural afflictions of hunger, climate, accidents, or parasites, to those supposedly caused by the supernatural interventions of devils, witches, or gods—but the basic view of sickness arose from our deepest anxieties, and our ministrations of healing developed in relation to religion and even magic.[2] Only recently has a strict scientific approach intervened in Western medicine, a development that has brought remarkable new healing power.

Perhaps the single greatest advance in the history of medicine, in our opinion, is the movement away from an approach to the sick that was characterized by shunning and abandonment, toward one in which the sufferer is fed, protected, and nurtured like a child. Without this most basic concept and covenant, nothing is medically possible. But this did not happen in one place or at one time.[3] And the path toward compassionate healthcare has been bumpy to say the least. One of the biggest potholes on the road to healing has also been one of its major motivations: money.

The evolution of disease, and of medical beliefs and practices to confront those diseases, has always reflected developing economic, political, and social complexities. With the advances in agriculture came population centers with increased numbers of people living in relative proximity. This allowed for the emergence of epidemics, so sick individuals could no longer be left behind. As societies became hierarchal, work became more specialized and it became clear that the loss of a particular individual's labor could weaken the group. From this came the realization that illness could spread, further hampering productivity and threatening society as a whole, thus sparking the devolpment of the healing arts. And as the idea of healing matured, so did the number of people with special interest and skill in this area. Shamans, midwives, herbalists, bonesetters, wise women, medicine men, barber-surgeons, and healer-priests all contributed to the traditions that developed into the physician and other healing specialists. Unfortunately, history also teaches us that corruption has not escaped the healing arts.

The Middle Ages may be the most obvious and destructive example of ecomonics interfering with medicine. The imposition of fee-for-service throughout Europe during this critical era was one of the most corrupting influences in the history of science. Fee-for-service medicine distorted the healing tradition in many ways but none more so than the wholesale destruction of female practitioners who were often burned at the stake as witches, rather than be allowed to compete with the highly profitable and church-sanctioned male practitioners. In fact, the economics of healing has had such a profound impact upon this benign practice that it must be examined in our consideration of medical advances.

The idea of healing is shared by virtually all cultures but it is a complex notion, and there is a dynamic equilibrium between opposing reactions to illness that moves back and forth even today, as we see in responses to people with AIDS. In some places sick people are taken in regardless of citizenship, employment status, membership in an insurance plan, or ability to pay. In other places the sufferer is left by the side of the road, even if the road is beside the door of a hospital.

Science and technology generally progress: more immunizations are developed, the CAT scans get better, magnetic resonance imaging (MRI) proves to be a valuble tool, and the genetic code is deciphered, which leads to gene therapy. But the greatest advances have to do with human values, and the greatest value lies in the concern of one for another. Still, we have seen that the quality of human values when it comes to healing ebb and flow. Consider those who today are living in degradation, for they show how we are still backward with respect to our response to suffering and our concern for the future. The idea of healing, and the values it encompasses, are tenuous and always under threat. In the dream of development we must hear all voices, lest we drown among the artifacts and the money. The Apollo of myth and legend heard the cries of the vulnerable and the suffering. He taught the healing arts to Chiron, and then on to Jason, Achilles, Athena, and Aesculapius. We must be mindful of that spirit.

Fig. 3. Hippocrates the physician (c. 460–377 BCE), who believed that disease had a natural cause and cure and was not sent by the gods. He believed balancing the body "humors" would maintain health.

2.

The Doctor-Patient Relationship

As to diseases, make a habit of two things—to help, or at least to do no harm. The art has three factors, the disease, the patient, the physician. The physician is the servant of the art. The patient must co-operate with the physician in combating the disease.

—Hippocrates, *Epidemics*, 1:11

The special relationship between the healer and the sick person is the origin and the heart of medicine.[4] It is the most remarkable and wonderful social contract in human history. Here, two strangers come together, away from the sight or knowledge of others. They are separated by the pain and suffering of illness, by dependency, by ideas and approaches to disease, and often by financial transactions. Still, they must forge a bond of intimacy and trust wherein the most personal and sensitive information is shared, along with the greatest fears, the deepest desires, the fondest hopes. When the relationship works, it can, by itself, have extraordinary powers of healing. For example, in the reassurance and comfort that can replace panic in the face of acute illness, or the confidence that can replace doubt during a long course of treatment, or the ability to withstand intolerable pain

25

Fig. 4. *Melencolia*, a 1514 engraving by Albrecht Durer, depicts one of the four humors illustrated in the inset chart—Flegmatic, Sanquine, Melancholy, and Coleric.

upon the arrival of the doula, midwife, or doctor. Modern technology is contributing advances that should strengthen communication and trust between doctor and patient. The Internet, with its instantly available information about illness and treatment, can enrich dialogue and understanding between the two.

The energy and restorative power in this unique bond is easily appreciated and has been valued in many healing traditions. Yet it provides power that can be misused with harmful results. As Florence Nightingale observed, medicine "is a work which makes angels or devils of men." The growth of modern scientific medicine and the introduction of money as a party to the relationship have shaken the foundations of this amazing advance in human interaction. Consider the evolution of our terminology. The relationship has gone from one between the *healer* and the *sick person* to that between the *doctor* and the *patient*. And while the doctor-patient relationship has maintained its benign mandate in the doctor's commitment to work for the best interests of the patient, the terms suggest a more distant, more complex, less personal bond. Scientific medicine relies on molecular mechanisms, specific treatments, and surgical cures, which should work regardless of any relationship. *Doctor* is more specialized, more set apart than *healer*. In theory, scientific medicine should lessen the requirement for a human bond. In practice, the bond has been weakened, but its diminished strength is to the disadvantage of both doctors and patients.

As treatments become more powerful, as potential for complications increase, as understanding of health and disease deepens, so does the need for communication and understanding between the healer and the sick person. Nevertheless, now even *doctor* and *patient* have been replaced with *provider* and *customer*. There is the deepest significance in these terms. The intrusion of insurance companies into this equation has distanced the relationship further and diluted the communication and bond between the healer and the sick person. Both doctor and patient must respect the inherent value in their relationship to each other in order for either to prosper. All the technology in the world will never change that.

3.

The Hippocratic Corpus

The Hippocratic Corpus refers to an extensive body of texts written between about 440 and 340 BCE that are ascribed to the legendary Hippocrates (460–377 BCE) and/or his many followers. It is a diverse work with both philosophical and practical treatises, but the core rests upon a rational view of the world in which health and illness can be studied and understood through careful observation and the application of reason, rather than by assuming and accepting the possibility of divine or other supernatural intervention. Through these works, disease is considered to have a natural cause; and this great advance was carried through Hippocrates' younger contemporary, Plato, and to Plato's student, Aristotle. Aristotle was the son of a doctor, and in addition to his many contributions to philosophy, ethics, politics, art, and cosmology, he extended the rational basis of medicine by introducing a scientific method and an investigative approach that form the basis of contemporary Western medicine (see advance 84).

The corpus contains works titled *On the Nature of Man; On Regimen; On the Sacred Disease; Epidemics;* and *Airs, Waters, Places*—among some sixty others. Within the strictures of this body of work, Hippocratic physicians sought to understand the general characteristics of racial groups and nationalities, as well as health and disease, in terms of balance or imbalance in the amount or flow of the bodily fluids (chymoi): blood, phlegm, yellow bile, and black bile. It was a generally conservative and restrained medicine, certainly in terms of therapeutics, where watchful waiting for return of the equilibrium of the fluxing fluids was encouraged, perhaps influenced by proper diet, exercise, rest, personal hygiene, and judicious sexual activity. bloodletting was certainly the most dramatic and long-lasting treatment of Hippocratic medicine. Endorsed by Galen (see advance 5), bloodletting persisted well into the twentieth century. The ability to predict the outcome of an illness, to prognosticate, was valued above therapeutic intervention—a seemingly rational orientation for physicians of that time. But the most enduring aspect of the Hippocratic Corpus is its instruction to the healer with regard to the fundamentals of professional life and conduct.

The Hippocratic Corpus, in many of its timeless aphorisms and especially in the Oath, represents the earliest and most important expression of correct professional conduct. The first principle of medicine, *primum non nocere*, "first and foremost do no harm," frequently mistaken as part of the Oath, is certainly the most important statement civilization has regarding the life and work of healers. And these instructions can be equally applied beyond the boundaries of medicine, as in the famous observation on the life of the physician, "Life is short, the art long, opportunity fleeting, experience fallacious, judgement difficult." And the Oath itself—notwithstanding its controversial approach to abortion, which, along with infanticide, was then common practice; its separation of medicine and surgery; and its clear exclusion of women—continues to define the high moral purpose of the healer.

The Hippocratic Oath

I swear by Apollo the physician, by Aesculapius, by Health and all the powers of healing, and call to witness all the gods and goddesses . . . to reckon him who taught me this Art equally dear to me as my parents, to share my substance with him, to relieve his necessities if required; to look upon his offspring in the same footing as my own brothers, and to teach them this art, if they shall wish to learn it, without fee or stipulation; and that by precept, lecture, and every other mode of instruction, I will impart a knowledge of the Art to my own sons, and those of my teachers, and to disciples bound by a stipulation and oath according to the law of medicine, but to none others. I will follow that system of regimen which, according to my ability and judgement, I consider for the benefit of my patients, and abstain from whatever is deleterious and mischievous. I will give no deadly medicine to any one if asked, nor suggest any such counsel; and in like manner I will not give to a woman a pessary to produce abortion. I will be chaste and religious in my life and in my practice. I will not cut, even for the stone, but will leave such procedures to the practitioners of that craft. Into whatever houses I enter, I will go into them for the benefit of the sick, and will abstain from every voluntary act of mischief and corruption; and, further from the seduction of females or males, of freemen and slaves. Whatever, in connection with my professional practice or not in connection with it, I see or hear, in the life of men, which ought not be spoken of abroad, I will not divulge, as reckoning that all such should be kept secret. If, therefore, I observe this Oath and do not violate it, may I prosper both in my life and in my profession, earning good repute among all men for all time. If I transgress and forswear this Oath, may my lot be otherwise.

Suggested Reading

1. Steven H. Miles, *The Hippocratic Oath and the Ethics of Medicine* (Oxford: Oxford University Press, 2005).

2. The *Journal of Medical Ethics*. See *JME* online at http://jme.bmjjournals.com/.

4.

The Discovery of
Microscopic Life

My work, which I've done for a long time, was not pursued
in order to gain the praise I now enjoy, but chiefly from a
craving after knowledge, which I notice resides in me more
than in most other men. And therewithal, whenever I
found out anything remarkable, I have thought it my duty
to put down my discovery on paper, so that all ingenious
people might be informed thereof.

—Antonie van Leeuwenhoek,
letter to the Royal Society (June 12, 1716)

Remember Leeuwenhoek! Antonie van Leeuwenhoek of
Delft, Netherlands, born in 1632, fabric merchant,
sometime surveyor and wine assayer, and Chamberlain
to the sheriffs of Delft, who—without higher education
but with passion for observation and description, and
using single lens microscopes of his own construction—discovered
"very little animalcules" in rainwater, in pond and well water, and in
his own bodily fluids and excretions. He opened the world of the
unseen, where microorganisms and human cells struggle in health and
disease.

It was the mid-seventeenth century, and Holland was in love with
commerce, music, art, and learning. There was a school in every vil-

lage, and as Descartes observed, "There is no country in which freedom is more complete, security greater, crime rarer, the simplicity of ancient manners more perfect than here." In 1648 sixteen-year-old Antonie was sent to Amsterdam upon his father's death, where he was apprenticed to a linen draper. There he became interested in using lenses to examine cloth. Upon returning to Delft in 1654 he opened a fabric business and, some years later, he learned to grind lenses. Leeuwenhoek's "microscopes" were actually fine magnifying glasses: single tiny lenses mounted in a brass plate with a screw to adjust the position of the specimen up and down, and another screw to focus by moving it away from or toward the lens. The instrument was a few inches long and was held close to the eye, requiring great patience to use. Still, he was able to achieve magnification of over two hundred times, with better images than anyone of his day because of the excellence of his lenses and a method of lighting his specimens for which the details have now been lost. Compound microscopes using two lenses had been invented in 1595 (see advance 30) well before Leeuwenhoek's birth, but difficulties in lighting the specimens and adjusting focal lengths limited their usefulness. It was Leeuwenhoek's curiosity that distinguished him rather than his instrumentation.

In 1673 Leeuwenhoek began a fifty-year correspondence with the new Royal Society of London with an observation on the stings of bees.[5] His letters, written in Dutch, were translated into English and Latin and published in the Philosophical Transactions of the Royal Society. These letters contain the first descriptions of bacteria, protozoa, the red blood cell, spermatozoa, the mouthparts of insects, the fine structure of plants, the striations in muscle cells, the life cycle of the flea, parthenogenesis, the uric acid crystals of tophacious gout, and much more. He challenged "spontaneous generation," the prevailing view that living things can develop from inert material.

His letter of September 7, 1674, described the green charophyte alga, spirogyra, in lake water as, "green streaks, spirally wound serpent-wise, and orderly arranged, after the manner of copper or tin worms . . . about the thickness of a hair . . . consist[ing] of very small green globules joined together." On September 17, 1683, he wrote of his observations of the plaque taken from between his own teeth and from an old man who never cleaned his mouth. There he had seen, "an unbelievable great company of living animalcules, a-swimming more nimbly than any I had ever seen. . . . The second sort . . . ofttimes spun round like a top. . . . The biggest sort . . . bent their bodies into curves . . . the other animalcules were in such enormous numbers, that all the water seemed to be alive." He described the first protozoa, *Giardia lamblia*, in his own stool, believing it to be the cause of his diarrhea some two hundred years before it was proven to cause disease. He hired an illustrator to draw what he described, and the pictures were remarkably accurate. The letters made him famous, and, in 1680, he was elected a member of the Royal Society, although he never attended a meeting. The mighty flocked to see Leeuwenhoek's microscopic worlds. In 1698 he showed the blood circulating in the capillaries of an eel to Peter the Great of Russia. Leeuwenhoek died on August 30, 1723.

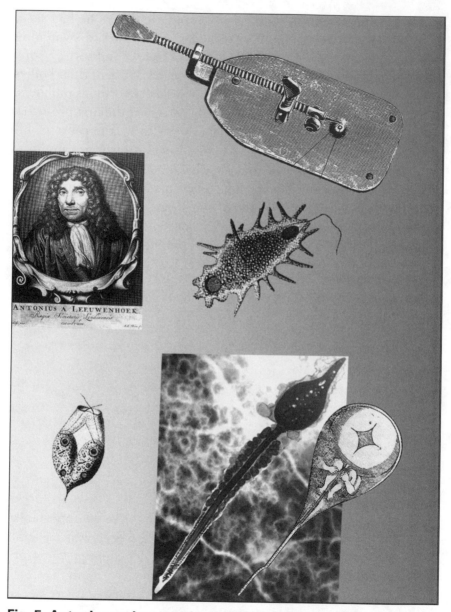

Fig. 5. Antonie van Leeuwenhoek (1632–1723) with his "micro-scope" and a few examples of his discoveries—(*left to right*) two protozoa and a contemporary image of a sperm next to a mythical image made in 1694 of a sperm within which can be seen a "humoncolus" (a cartoon of a miniature human).

5.

The Medical Encounter and the History of the Present Illness

History takes time . . .

—Gertrude Stein (1874–1946)

No instrument, no image, no diagnostic test has the power to reveal the nature of illness and a specific diagnosis as swiftly, safely, painlessly, and accurately as the medical history. It is generally agreed that a well-crafted medical history can yield the correct diagnosis in more than 80 percent of cases. Of course, the healer must be skillful and knowledgeable enough to obtain and precisely understand the sick person's complaints, feelings, and observations.

The medical encounter, as old as healing itself, has evolved in the Western tradition to become a complex tool of unrivaled importance. Yet, until the last thirty years or so, very little had been written about it in the medical or pedagogic literature. The encounter, and especially the first encounter, has a standard format consisting of the medical history and the physical examination. Here we focus on the medical history, which is, by far, the most powerful tool we have for establishing trust and respect, and for making a diagnosis.

Through the medical encounter—before, during, and after working

35

together on the medical history—the healer and the sick person must begin to forge their relationship. The history itself is divided into two specifically crafted parts: the History of the Present Illness (HPI), which deals with the patient's major immediate complaints within the context of the process that led to these symptoms, and the Standard Data Base (DB), which deals with old problems, background health information, and coexisting minor problems. The HPI begins with the Chief Complaint (CC), a single sentence, using the patient's own words, stating the symptom(s) that brought the illness to attention or the reason for the visit. This is followed by the narrative body of the HPI. The DB is divided into the HPI, the Past Medical History (PMH), the Social History (SH), the Family History (FH), and the Review of Systems (ROS). With these data, the doctor is prepared to perform a complete, yet focused, physical examination, decide which laboratory tests should be obtained, and consider specific diagnoses and treatments.

Among these components of the medical encounter, the HPI is the most vital. Obtaining accurate data, analyzing these bits of information, and constructing a narrative of an individual's health and disease history is the most difficult cognitive process any healer must face. In the training of physicians, struggling with the HPI takes center stage early, and honing this tool becomes a lifetime enterprise. It is the HPI that represents one of the very greatest advances in the history of medicine. It appears deceptively simple, is highly stylized, yet leaves room for a world of nuance. Nothing can distinguish a superior physician as well as her HPI of a confusing case. Here is a brief and simple example:

(CC) Mr. Jones is a 60-year-old man who has been having "viscious chest pains" for the past week.

(HPI) John Jones was found to have high blood pressure nearly twenty years ago, but has taken no medication for this during the past five years. Ten years ago he was diagnosed with diabetes and since then has taken one pill each morning of an unknown (can't remember) medication to control his blood sugar. Little specific information is available regarding his blood pressure or diabetic control, but he does report frequent occipital headaches in the morning, and frequent nocturnal urination. His father was hypertensive and died of a stroke at age 55. His mother is a Type 2 diabetic. An older brother died of a heart attack at age 43. Seven days ago, Mr. Jones experienced chest pain for the first time while sitting in a chair watching TV. The pain felt like "someone sitting on my chest," and lasted about one hour. It did not radiate, and there was no sweating, nausea, or vomiting, and he had no shortness of breath or swelling of his legs. Since his first episode of pain, he has had daily chest pain whenever he climbs the flight of stairs to his apartment. The pain subsides within one minute of his reaching the top of the stairs. He has never had an electrocardiogram.

With this HPI, a working diagnosis of coronary artery disease can be made. The location and character of his pain are compatible with at least a half-dozen diagnoses. But John has a family history of cardiovascular disease and at least two major predisposing factors for coronary artery disease. And the details of the initiation and duration of the pain suggest that the first episode was a myocardial infarction ("heart attack" in which muscle dies), while the subsequent episodes were those of angina pectoris (painfully insufficient coronary blood flow not necessarily causing death of heart muscle). In a reasonably short time, an organized professional using the highly refined techniques of the HPI can make this and most other diagnoses. A complete but focused physical examination can then be conducted, and a plan for diagnostic tests and treatment can be formulated. These tools are complementary, and each patient deserves the benefits of all appropriate methods.

Suggested Reading

1. Lynn S. Bickley and Peter G. Szilagyi, *Bates' Guide to Physical Examination and History Taking* (bk and CD), 7th ed. (Philadelphia: Lippincott, 1999); 8th edition (2003).

2. Richard F. LeBlond, Richard L. DeGowin, and Donald D. Brown, *DeGowin's Diagnostic Examination*, 8th ed. (New York: McGraw-Hill Medical, 2004).

3. Henry M. Seidel et al., eds., *Mosby's Guide to Physical Examination*, 5th ed. (St. Louis: Mosby, 2003).

Fig. 6. Claudius Galen (c. 131–200 CE) with an early diagram of the circulatory system as he understood it and, below, a depiction of him attending to a fallen gladiator.

6.

Galenic Medicine

laudius Galen, born in Asia Minor in 131 CE, began studying medicine at the age of seventeen and produced a body of work that represented the most important medical scholarship for nearly sixteen hundred years. He lived about seventy years and devoted many volumes of his work to the philosophy of the spirit, the four humors, remedies and pharmaceuticals, and anatomy and physiology.

In his youth, Galen traveled widely through Greece, Asia Minor, Egypt, and Palestine; read Aristotle, Plato, and Hippocrates; and, at twenty-eight, became physician to the sanctuary of Asklepios in his home town of Pergamum, Greece. Here he was spiritually close to the source of Western medicine. In Greek mythology, Asklepios was the son of Apollo the physician and carried the symbol of medicine known as the caduceus, or the staff of Asklepios. After five years Galen moved to Rome, quickly gained fame as a teacher of medicine, and became physician to the emperor Marcus Aurelius. But Galen was a monotheist whose works refer to Christ and Moses. While his ideas overthrew some of the teachings of Aristotle and Hippocrates, his ideas became so entrenched and followed that it became very unhealthy to question his thinking.

Hippocrates had developed the notion of "humoral pathology," or the idea that health is dependent upon the proper balance of the four "humors," or body fluids (blood, phlegm, yellow bile, and black bile (see advance 3). Galen accepted the core of this belief and suggested that human personality was also determined by the relative predominance of one of these. Thus, an individual was sanguineous, phlegmatic, melancholic, or choleric, because of the humors—an early consideration that physical constitution has influence upon personality.

Galen was a busy doctor who established his own extensive pharmacy with hundreds of medicines made from animal and vegetable ingredients. He was one of the founders of pharmacological science, for which his major contribution was the concept of dosage. He prescribed precise quantities of each ingredient, and some of his medicines contained more than twenty ingredients. He was convinced that incorrect dosage could result in little effect, great effect, harm, or even death. But his most significant contribution to medicine was not pharmacological—it was anatomical, in his relating the structure of organs to their function.

The mystery of the functions of organs was a central issue in medicine both before and after Galen's time. But Galen, through dissection and careful observation, made discoveries that advanced medicine more than anyone before in the Western tradition. His findings were many and mainly involved structure-function considerations of the skeletal system, muscles, nerves, and lungs. He is revered for delineating the relationship between paralysis and severance of the spinal cord. But equally important were his discoveries in the area of the circulation of the blood.

Aristotle had taught that the heart housed the soul and was responsible for thinking. Thus, before Galen, it was commonly believed that inspired air entered the arteries, which had been filled with nothing but air. Galen taught that thinking was the province of the brain, that food entered the blood through the liver, that blood from the liver nourished the other organs, and that the heart circulated blood through the arteries. These concepts were great leaps forward. However, he did not understand that blood circulates in one direction through a closed system. Major advances in anatomy and the physiology of blood flow awaited Adreus Vesalius in 1543 (see advance 7) and William Harvey in 1628 (see advance 10).

In the meantime, religious concerns thwarted the advance of science and medicine. It is not clear whether Galen did human dissections, which were forbidden in Greece for religious reasons. He may have done them in Egypt, but most scholars think not. He certainly dissected animals, including monkeys; however, his job as physician to the gladiators in Rome gave him ample opportunity to observe virtually all human organs even during life. The study of human anatomy was essentially banned for nearly fifteen centuries. Galen's authority was thought to be divine, and in the Middle Ages church officials referred to him as Divinus Galenus. In spite of the master's admonition to see for oneself, many who challenged his theories were burned at the stake.

7.

On the Fabric
of the Human Body
by Andreas Vesalius

In 1543, about 1,343 years after Galen's death, Andreas Vesalius—a Flemish physician and anatomist at the University of Padua—with great care and attention to scientific, artistic, typographic, and all other details, published his *De humani corporis fabrica, libri septem*, and thereby jump-started medicine out of its thirteen centuries of doldrums.

In the Middle Ages of Western culture, many thousands of "wise women" were burned as witches, many of whom were peasant healers whose empiricism challenged the "revealed" wisdom of religious and other rulers who had claimed divine right. The "devineification" of Galen was a crude tool for the control of people's thoughts and ideas. We take Andreas Vesalius, and his *Fabrica*, to represent the triumph of truth over dogma, the grand reentrance of reason, and the expulsion of religion from medicine.

It is not for the *Fabrica*'s magnificent works of art that we celebrate

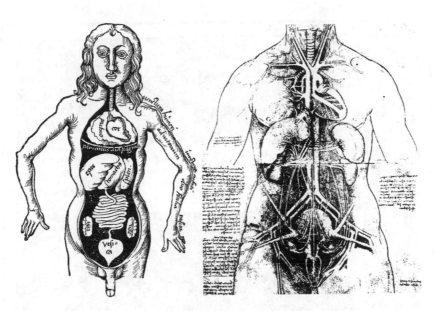

Fig. 7. *Left*—Medieval anatomical drawing of the human anatomy, 1503. *Right*—Anatomical drawing of the torso by Leonardo da Vinci, 1452–1519.

Vesalius. He did not paint them, and Michelangelo, Leonardo da Vinci, Albrecht Durer, and others before had established the relationship of art and artists with anatomy and dissection.

It isn't the perfection of his anatomy that elevates him to the point where many have considered the *Fabrica* to be medicine's crowning achievement and, like William Osler, called it "the greatest book in medicine." Book 1, devoted to the skeleton, and book 2, on the muscles, are astonishing in their anatomic detail and depiction. But books 3 through 7, dealing with the veins and arteries, the nervous system, the abdominal organs, the heart and lungs, and the brain, are less meticulous and contain striking errors and omissions. He missed the pancreas, the adrenal glands, and the ovaries. He did a poor job with the female sexual organs.

It is because Vesalius challenged Galen that we revere him. And it is because he challenged Galen out of reverence for him. The *Fabrica* was a breakthrough series of books because it brought freedom to

Fig. 8. Title page of the first edition of *De Humani Corporis Fabrica* (1543) by Andreas Vesalius.

study withour fear of oppression. In Vesalius's time medical people studied human anatomy in Padua, Pavia, Bologna, and Paris, the few places that would allow it, with the professor perched on a high chair, far from the corpse, as he read to his students from Galenic texts. A barber did the cutting, and no one looked to see that the heart did not have three ventricles, as the text maintained, or that the uterus is not multichambered, or that the lower jaw is made of one bone not two.

Vesalius followed the Galen who said that you must see for yourself. When necessary, he fought packs of dogs to obtain dead bodies from cemeteries, stole them from gibbets, or received them as gifts after public executions. He would "keep them in [his] bedroom for several weeks" while he dissected them—unpreserved, diseased, rotting bodies. Vesalius saw for himself. He described what he saw. And this was not a simple matter as the form and power of the modern state was evolving and requiring methods for the control of ideas. But Vesalius was driven, and what he produced was so extensive, so grand in design, detail, and illustration, that its value was obvious despite its being denounced by conservative contemporaries as arrogant, ignorant, impious, poisonous, and worse. As a result, anatomy, physiology, and all medicine advanced, free of the bonds imposed by Divinus Galenus.

What predominantly drove the man was ambition—not lust for knowledge, or scientific freedom, or understanding that for culture to advance creative intelligence had to be thrown into the mix. His father had been pharmacist to Emperor Charles V. Vesalius simply wanted to be physician to the emperor, or so most students of medical history believe. He carried his great work over the Alps to Basel to be printed on the finest papers. He watched over the printing process so that the wonderful woodcut engravings would receive proper attention. In 1543 he presented his masterpiece to Charles V—a large, seven-hundred-page volume bound in purple silk velvet—with hand-painted illustrations. His goal was that it would land him a job. Within weeks he was invited into the imperial service. He produced nothing further of value.

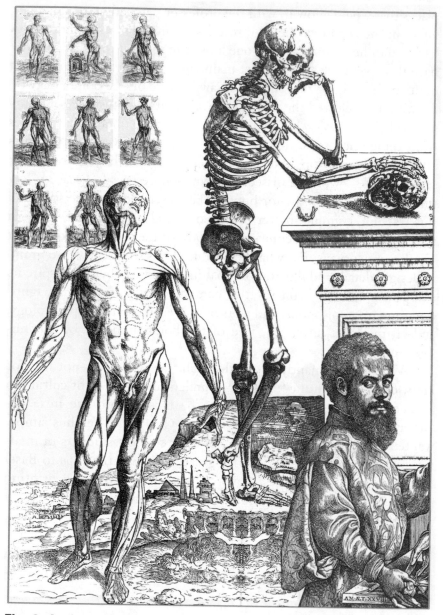

Fig. 9. Andreas Vesalius (1514–1564) (*lower right*) with some illustrations from his *Fabrica*.

8.

The Physical Examination

God bless the physician who warms the speculum or holds
your hand and looks into your eyes. Perhaps one subtext
of the health care debate is a yen to be treated like a whole
person, not just an eye, an ear, a nose or a throat. A yen to
be human again, on the part of patient and doctor alike.

—Anna Quindlen (b. 1952), journalist

Observation, palpation, percussion, and auscultation are
the fundamental techniques of physical examination.
The concepts and methods have developed since the
time of Hippocrates, but have they been refined and
systematically applied by physicians during the past two
hundred years. They employ some of the most sensitive and precise
tools imaginable: the human senses. And they serve remarkably well.

The physical examination is conducted after information is
obtained about the patient's history, and so the examiner generally has
a good idea of what problems may exist (see advance 5). Nevertheless,
during the initial encounter a complete physical examination should
be done in a meticulous and systematic fashion. It begins with a general observation of the patient's appearance. Does she appear to be her
stated age? Does she look anemic, jaundiced, poorly nourished? Next,
the vital signs of temperature, blood pressure, pulse, and respiratory
rate are measured and recorded. The examiner then proceeds with
specialized examinations of the head and neck, the chest, the breasts,

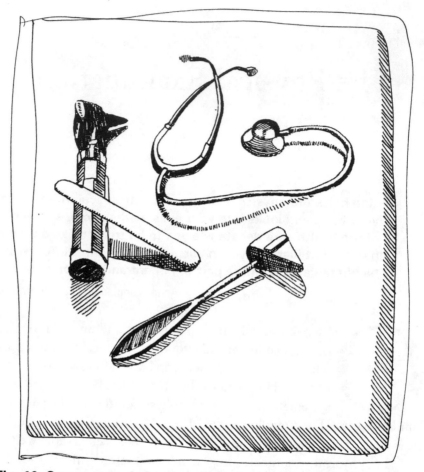

Fig. 10. Common tools for the physical examination—(*left to right*) tongue depressor, otoscope, reflex hammer, stethoscope.

the lungs, the cardiovascular system, the back, the abdomen, the rectum, the sexual organs, the extremities, and the nervous system, including detailed assessment of the state of consciousness and aware-ness. While description of the precise techniques used to evaluate these systems is beyond the scope of this chapter, the general method-ology involves observation, palpation, percussion, and auscultation, applied in that order. We will briefly describe these methods.

Observation requires knowledge of the normal state and the

ability to describe and record abnormalities. Are the membranes of the eye and mouth pink or pale? Are the pupils round, regular, and equal in size? Do they react by constricting when a light is directed at them? Is the abdomen flat or distended? Are the fingers distorted by swollen joints? Are they red, deviated laterally?

Palpation, or feeling with the hands, puts the healer in direct physical contact with the patient and thus requires great care and sensitivity. Is there a lump in the breast? What are its dimensions? Is it hard, tender, or fixed to the underlying chest wall? Are there other lumps in the breast, under the arm? Is the abdomen tender? Can the liver or spleen be felt? There are many ways of positioning the body, using the movements of breathing, and other methods of enhancing the ability of the examiner to feel normal and abnormal structures. We must appreciate structure because it determines function.

By tapping over structures with the fingers, the examiner may appreciate their size, shape, and consistency based on the sound and feel that is produced. This is the art of percussion, and it is particularly useful over the chest where the size and shape of the heart and the presence of fluid or consolidation in the normally resonant air-filled lungs can be determined with accuracy. Over the abdomen, the size and shape of an enlarged liver or spleen may be revealed, and distention can easily be found to be caused by a gas-filled intestine or fluid in the peritoneal (abdominal) cavity.

Auscultation, or listening to the sounds that come from the body, was first described by Hippocrates, who simply placed his ear directly on a patient's chest. But it was not until the great French physician Theophile Hyacinth Laennec (1781–1826) invented the stethoscope that auscultation began to develop as a major adjunct to the physical examination. It is said that Laennec was inspired when he watched two children playing a game with a nail and a long beam. As one end of the beam was tapped with a nail, the child at the other end listened for the signals transmitted through the wood. Laennec first rolled a piece of paper into a cylinder and, placing it between his ear and a patient's chest, heard the heart sounds clearly. He then fashioned stethoscopes of wooden cylinders. The modern biauricular stethoscope, with its ear pieces connected by rubber tubing to a bell and diaphragm, evolved in a number of steps, though an Austrian physician, Joesf Skoda, is credited with the major innovations. By careful auscultation the skilled examiner can precisely determine normal function or a vast array of structural abnormalities, particularly in the heart, lungs, and intestines as blood, air, or intestinal contents rushing through these organs produce characteristic sounds.

The physical examination involves touching and feeling—intimate contact with virtually every part of the body. This provides an opportunity to extend the bond of trust begun during the sharing of information in the history taking. It must be performed with the greatest of tact and sensitivity. While it may strengthen a relationship that is fundamental to the healing process, it is clear that abuse of this contact is an unforgivable transgression. It is also true that the physical examination, like the history, takes time and great skill, commodities which are considered costly in today's market. In addition, technological advances such as CAT scans, magnetic resonance images, and laboratory tests can provide information that can clarify many diagnoses.

Managed care organizations find it profitable to reduce the numbers of physicians by restricting the time spent in history taking and the physical examination. Then they limit the batteries of scans and tests that physicians use to try to compensate. The errors and abuses resulting from this approach are obvious to experienced physicians, but they are not considered in the cost analyses that drive a system that responds to the "bottom line."[6]

Only a decade ago we still taught medical students and young physicians to "spend some time, get to know your patients, be meticulous with your histories and physical examinations." That's old fashioned now. Primarily for economic reasons, the physical examination is currently in retreat. If we continue to be ruled by the "profit-over-all" approach to healthcare, the physical examination will soon be extinct. This will be a great loss, and the trend in that direction has already resulted in errors and suffering.

9.

To See for Oneself
at the Autopsy

Fig. 11. Medieval depiction of a postmortem dissection.

Vesalius (see advance 7), and others before him (see advance 6), performed postmortem examinations, but it took another century before doctors realized the full medical potential of the recently deceased. Vesalius and his ilk were motivated by passionate interest in pure anatomy. They understood that there is a relationship between structure and function. And while the early anatomists were interested in description as it related to normal function, it took time for a clear progression to the pathologist, who studies pathologic anatomy to understand disease. The legacy of Vesalius includes the necropsy studies of Giovanni Battista Morgagni (1682–1771). Morgagni, like Vesalius before him, was professor of anatomy at Padua. He performed over seven hundred autopsies and published *De sedibus et causis morborum* (1761) (On the Sites and Causes of Disease).

The introduction of routine autopsy to relate the signs and symptoms of disease to morbid anatomy, and to determine the cause of death, dominated the advances in medicine for nearly two hundred fifty years. The central role of the autopsy in medical science was best expressed by its most prolific practitioner, Carl von Rokitansky (1804–1878), who performed some sixty thousand autopsies. Rokitansky developed a method of dissection in which most of the viscera are removed together in a rapid and efficient fashion. His *Hanbuck der pathologische Anatomie* (1842–1846) (Handbook of Pathologic Anatomy) made notable contributions in the areas of heart disease, congenital malformations, peptic ulcers, and pneumonia, among others. Rokitansky proclaimed that "[p]athological anatomy must constitute the groundwork, not only of all medical knowledge, but also of all medical treatment. It embraces all that medicine has to offer of positive knowledge."[7]

From our current perspective, Rokitansky's statement seems excessive. But his view predominated well into the second half of the twentieth century. Of course, in the nineteenth and twentieth centuries autopsies added biochemistry, toxicology, electron microscopy, and other methods to basic gross anatomical and light microscopic analysis. Nonetheless, today we see virtually all hospital autopsy rates falling to much lower levels. Much of this can be explained by the fact that CAT scans, PET scans, MRI, ultrasound, and other methods allow accurate premortem diagnosis. Still, the autopsy method had such a long run as medicine's central tool that now, as it declines, there is anxiety at its passing.

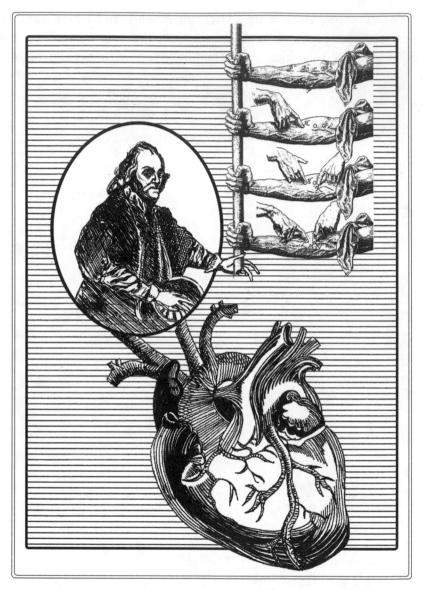

Fig. 12. William Harvey (1578–1637) holding his sketch of his test that applied and released pressure to veins, showing they emptied, then refilled with blood, and demonstrating that valves in veins allow blood to flow in one direction only. He proved that blood circulated through arteries and veins, pumped by the heart, which he published in his book *De Motu Cordis* (1628).

10.

On the Motion of the Heart by William Harvey

P ut your ear to someone's chest (see advance 8), even your dog's will do, and listen to the familiar *lub-dub, lub-dub, lub-dub* of the organ that has inspired philosophers, poets, lovers, scientists, money lenders, and even politicians throughout the ages. As in watching tropical fish, this activity may lower your blood pressure, but consider how much time went by before humans fully grasped the concept of the circulation of the blood and the pumping action of the heart. Surely, there was always interest and speculation about this most obviously dynamic of internal organs.

Hippocrates (see advance 3) wrote, "The vessels which spread themselves over the whole body, filling it with spirit, juice and motion, are all of them but branches of an original vessel. . . . I protest I know not where it begins or where it ends for in a circle there is neither beginning nor end." He used dissection as his method, and discerned that form and function are related. He came very close to a working description of circulation, but it lacked crucial details. Where did the vital gases come from and go to? In what directions did blood flow in the great vessels coming directly from the heart? How does blood get from the right side of the heart to the left side? Does blood get propelled by the squeezing action of the heart's contraction (systole), or by the sucking action of relaxation (diastole)? Do the arteries, through their pulsation, pump the blood? What is the function of the valves within the veins?

Of course Galen (see advance 6) weighed in on many of those central questions. He thought that blood traveled from right side to left side through invisible pores in the wall (septum) that separates them. He believed that there was a two-way flow between the lungs and the right side of the heart through the pulmonary vein, with air going from lung to heart, and "sooty vapours" going back in the other direction. These notions were not supported by careful and accurate observations, and they were incorrect, but Galen was addressing the questions and making progress. And it was not his fault that authoritarian reactionary forces would stop further progress for so many centuries to come and even cause the culture to lose much of the knowledge and understanding that it had painstakingly acquired. The struggle between progress and reaction is always active. We maintain that this is the most relevant and meaningful lesson in these stories: there are always forces trying to quash advances, trying to own them, trying to turn them into cash, and trying to use them to separate and control people.

In England, as in all of Europe, the Dark Ages succeeded in commercializing medicine, placing it under the authoritarian monopoly control by the church and state, and putting treatments beyond the means of the poor. With the Enlightenment came reform in healthcare, poverty relief, and education. Nicholas Culpeper (1616–1654) struggled against the authoritarian monopoly of the Royal College of Physicians and tried to bring affordable treatments to London's poor. He translated the college's Latin *Pharmacopoeia* into English, and then, after vicious reprisals by the college, he wrote *The English Physician Enlarged* (1653), which provided affordable remedies for hundreds of human afflictions. The book went through scores of editions, and it is still available as *Culpepper's Herbal* to this day.

William Harvey (1578–1657) was a feisty fellow who called his detractors "shitt-breeches" and always packed a shiv in his belt. Like virtually all medical innovators, he combined knowledge and understanding of what had come before with an inquiring mind. Educated at Cambridge, he then went to Padua in 1600 to study medicine. At that time anatomy still dominated medicine and Padua was the place to be (see advance 7). There, neo-Aristotelian thinking was at war with Galenic authority, and Harvey, on the side of the new-wave rationalists, took up the problem of the cardiovascular system.

Returning to London, Harvey applied direction and "eyesight inspection" (alas, he ignored the microscope, see advance 30) to determine that the heart functioned as a muscle, ejecting blood by ventricular contraction in systole; that arteries do not pulsate on their own but by the shock waves generated by ventricular systole; that the valves in the heart and veins are arranged so that blood can flow in only one direction; and that venous blood flows through the right auricle to the right ventricle, where it is pumped to the lungs via the pulmonary artery, and returned to the heart's left auricle via the pulmonary vein, from which it flows to the left ventricle to be pumped into the aorta.[8] He applied quantitative experimental studies, using frogs and other animals. Measuring the tremendous volume of blood ejected by the heart each minute and hour, he rejected Galen's idea that the blood was absorbed by the body and continually being replaced by blood being made in the liver. Thus, he proved that blood is in constant circulation. And it is this concept that is so essential to our understanding and practice of medicine today.

Harvey published most of his findings between 1620 and 1628.[9] During that period he wrote,

> But what remains to be said upon the quantity and source of the blood which passes, is of a character so novel and unheard of that I not only fear injury to myself from the envy of a few, but I tremble lest I have mankind at large for my enemies, so much doth wont and custom become a second nature. Doctrine once sown strikes deeply its root, and respect for authority influences all men. Still the die is cast, and my trust is in my love of truth, and in the candour of cultivated minds.[10]

It was the birth of cardiology, and much more.

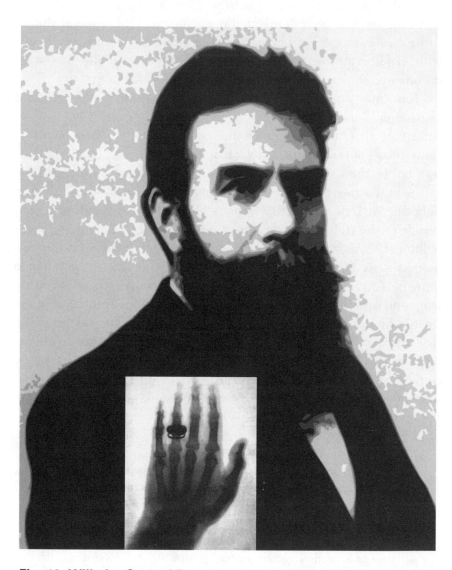

Fig. 13. Wilhelm Conrad Roentgen (1845–1923) with his x–ray of his wife Bertha's left hand and ring.

11.

Diagnostic X-ray Examinations

Much of what the healer seeks is hidden. The vital structures are covered by skin, muscle, and bone. For millennia our examinations of the internal organs were limited to feeling or listening for clues in their shapes and movements.

In the second half of the nineteenth century, ingenious instruments were devised to peer into those organs. The ophthalmoscope, the otoscope, the esophagoscope, the gastroscope, and the proctoscope are marvelous inventions that allow the viewing of organs that were previously beyond the powers of examination. Through the use of these tools, all developed between 1850 and 1895, we began to see inside. And while these glimpses were limited by proximity to the scope's portal of entry, they nonetheless produced revelations, fabulously expanding our ability to diagnose and treat. With continuing refinements those same tools are still essential after more than one hundred years. But the best was yet to come.

In 1895, with a stroke of near magic, came the power to see right through the body and examine virtually any structure. On November 8 Wilhelm Conrad Roentgen (1845–1923), a fifty-year-old professor of physics at the University of Wurzberg, Germany, was experimenting with a cathode ray tube. Roentgen was a brilliant experimentalist who had made important observations working with crystals, gases, and polarized light. Like many physicists at that time, he was fascinated with cathode ray tubes. These are simply sealed glass tubes that are evacuated so that they hold virtually no air, but contain electrodes through which electricity can be applied. With sufficient voltage the tubes glow and emit what are called cathode rays. The glow of the rays was observed to pass through the glass and can be seen a few centimeters around the tube. The mysterious glowing cathode rays were later found to be streams of electrons.

But on November 8, Roentgen wanted to see if there were invisible rays passing through the glass. So he enclosed the tube in black cardboard and darkened his laboratory. When he fired up the tube, he was amazed to find that a fluorescent screen that was placed more than a yard away was glowing a faint green. Invisible rays were passing through the glass tube and through the cardboard and bombarding the screen. He tried to block the rays with a playing card, a book, a piece of wood, rubber, and a thin sheet of aluminum, but still they passed right through. Then, placing a small lead disk between his thumb and forefinger, he held it before the screen and saw the shadow of the disk grasped in the bones of his hand.[11]

"I was as if in a state of shock," he would later say. For weeks he was grumpy and spoke little; rarely leaving the lab, he worked feverishly. "I have discovered something, but I don't know whether or not my observations are correct," was all the shy, self-doubting scientist would tell a friend. Then, on December 22, 1895, he took his wife, Bertha, into the lab and showed her his work, taking the first x-ray photograph of the human body by placing her hand on a photographic plate and exposing it to the mysterious radiation. She was frightened, experiencing the sight of her skeleton as a premonition of death.

Roentgen published his findings in the *Journal of the Physical-Medical Society of Wurzburg* in 1896. The paper was titled "Eine neue Art von Strahlen" (A New Kind of Ray), and, as he predicted, it was a sensation. "Now the devil will have to be paid," he told Bertha as he mailed off the manuscript. Never before, or since, has a discovery with medical significance had such a rapid and powerful impact around the world. Within one year 49 books and 1,044 papers on x-rays were published. And while there were frivolous, even comic, aspects to the early development of x-rays, like radio-opaque panties to protect virtue from the prying eyes of imagined x-ray peepers, Roentgen and others instantly applied the discovery to medical diagnosis. And Roentgen's observation, that prolonged exposure caused ulcerating dermatitis, quickly led to therapeutic applications.

The discovery of x-rays inspired Henri Becquerel (1853–1908) and Marie Curie (1859–1934) to discover natural radioactivity in December of 1898. In 1901 Roentgen was awarded the first Nobel Prize for Physics.

Fig. 14. William Beaumont peering directly into the stomach and sampling gastric juice from Alexis St. Martin.

12.

Observations and Experiments on the Gastric Juice and the Physiology of Digestion by William Beaumont

In 1806 a wealthy farmer's son refused his father's offer of land in Lebanon, Connecticut, and instead took about one hundred dollars in gold and a good mare to begin wending his way north and west until he settled in the town of Champlain on the New York shore of Lake Champlain. There he worked as a teacher and, because teachers didn't make much money, he also worked as a clerk in a dry goods store. After about a year, he somehow got the notion of becoming a doctor, so he took the ferry across the lake to Vermont and apprenticed himself to a Dr. Benjamin Chandler. In two years he was Dr. William Beaumont (no medical school was required), and he joined the Army Medical Service. We will not dwell upon how he got in trouble with a superior, nearly fought a duel, quit the army, started a practice in Plattsburg, New York, then rejoined the

Army Medical Service in 1820 and wound up at Fort Mackinac on the Upper Peninsula of Michigan. There, on June 6, 1822, he was summoned to treat Alexis St. Martin, a nineteen-year-old French Canadian trapper who had been wounded at close range by a shotgun blast.

The shot had blown out a portion of St. Martin's abdominal wall below the left breast and caused a perforation of the anterior wall of the stomach. It took a year for the wound to heal, but the aperture in the abdominal wall never closed and, by moving the flap of skin that covered the hole, Beaumont could see inside and actually view and sample all the activities occurring within St. Martin's stomach (a gastric fistula). For a decade Beaumont, by his own account "a simple Army surgeon," chased and cajoled the reluctant St. Martin to study the motions, secretions, fluxes of mucosal blood flow, and digestive activities of the living stomach. Several times Beaumont hired detectives to track down St. Martin, finally getting him an army commission so that he could keep him close at hand. In 1833 he published *Experiments and Observations on the Gastric Juice and the Physiology of Digestion.*[12] The work is divided into seven sections: Of Aliment; Of Hunger and Thirst; Of Satisfaction and Satiety; Of Mastication, Insalivation, and Deglutition; Of Digestion by the Gastric Juice; Of the Appearance of the Villous Coat, and the Motions of the Stomach; and Of Chylification and Uses of the Bile and Pancreatic Juice. It concludes with fifty-one "Inferences from the Foregoing Experiments and Observations." Here Beaumont describes the stomach at rest and at work: its secretions and their digestive activity, its motions during filling and emptying, the waxing and waning of blood flow to the surface mucosal lining, and the appearance of damage caused by alcohol and other noxious substances. The book was a sensation, the first piece of good Yankee science—it challenged physiologists to consider the mechanisms through which the complex digestive processes were regulated, and it set an agenda for physiologists and physicians that has endured to this day.

13.

"Nervism" and the Integration of the Organ Systems

Oh the nerves, the nerves; the mysteries of this machine called man!

—Charles Dickens (1812–1870)

L ike Beaumont, another village lad would reject his father's path and later shake the worlds of physiology and medicine. Ivan Pavlov (1849–1936), the son of a Russian priest, was on his way into the family business when he was led away by the "progressive" (his word) ideas of D. I. Pisarev, the most eminent of the Russian literary critics of the 1860s, and I. M. Sechenov, the father of Russian physiology. Pavlov abandoned his religious career and decided to devote his life to science. He became a doctor, studied surgery, and settled on the physiology of digestion, no doubt influenced by Beaumont's book since he read widely. Moreover, at the 1977 Nobel Prize ceremony an elderly gentleman who worked in Pavlov's laboratory assured me that Pavlov had read Beaumont.

In beginning his 1904 Nobel lecture, Pavlov said, "It is not accidental that all phenomena of human life are dominated by the search for daily bread—the oldest link connecting all living things, man included, with the surrounding nature."[13] Here we see that he was certainly also influenced by Charles Darwin.

He continued by saying,

> Food finding its way into the organism where it undergoes certain
> changes—is decomposed, enters into new combinations and again
> dissociates—represents the process of life in all its fullness, from
> such elementary physical properties of the organism as weight,
> inertia, etc., all the way to the highest manifestations of human
> nature. Precise knowledge of what happens to the food entering the
> organism must be the subject of ideal physiology, the physiology of
> the future. Present-day physiology can but engage in the continuous
> accumulation of material for the achievement of this distant aim.[14]

Thus, acknowledging a debt to Beaumont, he went on to discuss his
extensive discoveries in the physiology of digestion in which he used fis-
tulas of his own invention, surgically created in dogs, which enabled the
functions of various organs to be observed continuously under relatively
normal conditions. In this manner he studied salivation, pancreatic
function, hunger, and satiation, and all in the living animal, quite as
Beaumont had studied St. Martin. Pavlov and his dogs became more
famous, and more productive, than Beaumont and the feisty trapper
with Pavlov's discovery of the "conditioned reflex" and its profound
influence on the development of psychology. He founded "behav-
iorism," extending his findings to mechanisms of the human mind.[15]

We are celebrating Pavlov for his emphasis on the regulation of
organ function. He was most interested in what integrated the activities of organs, a theme that ran throughout all of his career. In 1883,
while still a student, he presented his doctor's thesis, *The Centrifugal
Nerves of the Heart*. In this work he developed his idea of "nervism"
and showed that there existed a basic pattern in the reflex regulation
of blood pressure and the activity of the circulatory organs. Later,
with extreme clarity, he showed that the nervous system played the
dominant part in regulating the digestive process and expanded the
theory of nervism to suggest that virtually all of the body's organ-
system integration was controlled through the actions of nerves. If he
had only discovered this, only opened the issue of integration and
control, it still would have been a great achievement. But he provided
the questions and the methods for the next leap forward.

If we pour into the stomach or directly into the intestine pure gastric juice, or simply the acid which it contains, or even some other acid, our gland [the pancreas] will begin to function just as vigorously, or even more vigorously, than in the case of the normal thyme passing from the stomach into the intestine. The profound significance of this unexpected fact is quite clear. . . . The gastric laboratory uses its protein ferment [enzyme] under an acid reaction. Different intestinal ferments, and among them, naturally, pancreatic ferments, cannot develop their activity in an acid medium. Hence, it is clear that the first task of the laboratory [Pavlov considered each organ to be a laboratory] is to provide the neutral or alkaline reaction necessary for its successful activity. These circumstances are effected by the above-mentioned interrelations, since the acid content of the stomach, as already stated, induces secretion of alkaline pancreatic juice (and the higher the acid content, the greater the secretion). Thus, the pancreatic juice acts above all as a solution of soda. . . . Thus, the purposeful relationship of phenomena is based on the specificity of the stimuli, that correspond to similarly specific reactions. But this by no means exhausts the subject. Now the following question should be raised: how does the given property of the object, the given stimulant, reach the glandular tissue itself, its cellular elements? The system of the organism, of its countless parts, is united into a single entity in two ways: by means of a specific tissue which exists solely for the purpose of maintaining interrelations, that is, the nervous tissue, and by means of body fluids bathing all body elements. These same intermediaries transmit our stimuli to the glandular tissue. We have thoroughly studied the first kind.[16]

(The second kind had recently been discovered and is the subject of the next advance.)

Pavlov concluded his 1904 Nobel lecture with these words:

Essentially only one thing in life interests us: our psychical constitution, the mechanism of which was and is wrapped in darkness. All human resources, art, religion, literature, philosophy and historical sciences, all of them join in bringing light in this darkness. But man has still another powerful resource: natural science with its strictly objective methods. This science, as we all know, is making huge progress every day. The facts and considerations which I have placed before you at the end of my lecture are one out of numerous attempts to employ a consistent, purely scientific method of thinking in the study of the mechanism of the highest manifestations of life in the dog, the representative of the animal kingdom that is man's best friend.

He was a revolutionary in spirit and in science, who valued precision and truth above personal needs. In his early days, Pavlov's relationship with the Bolsheviks after the 1917 Revolution was conflicted. Early on the Bolsheviks valued the inheritance of science from the old order, which had created a scientific community superior to many in western Europe. But the situation in Petrograd became increasingly desperate and by 1921–1922, unable to work, Pavlov requested permission from Lenin to transfer his laboratory abroad. Lenin denied the request, and Pavlov refused privileges offered to him while his staff worked in conditions of near starvation. After returning from a visit to the United States in 1923, Pavlov publicly denounced the revolution, saying, "For the kind of social experiment that you are making, I would not sacrifice a frog's hind legs!" With the emergence of Stalin in 1924, Pavlov resigned his posts, saying, "I also am the son of a priest, and if you expel the others I will go too!" In 1927 he wrote to Stalin, "On account of what you are doing to the Russian intelligentsia—demoralizing, annihilating, depraving them—I am ashamed to be called a Russian!" Later on, with Russia under attack by Hitler, Pavlov moderated his criticism and continued his scientific work.

Ivan Pavlov died in 1936, in what was then called Leningrad. He was a humanitarian, a visionary scientist, an early animal rights advocate, and a dissident who was too revered to be touched by tyrants.

14.

"Chemical Messengers" and the Discovery of Hormones

The great nineteenth-century experimentalist and physiologist Claude Bernard (see advance 84) proclaimed that "[t]he stability of the internal environment is the prime requirement for free, independent existence."[17] He had shown in the latter half of the nineteenth century that the liver was able to elaborate and secrete sugar directly into the bloodstream. The regulation of normal concentrations of sodium, chloride, potassium, and other essential materials in blood and urine, and within cells, and the maintenance of body temperature, metabolic rates, and the coordination of organ system activities, especially in the digestive system, were becoming central issues for thought and study. It had long been speculated that the thyroid gland secreted some essential substance into the blood. With the development of experimentalism, scientists understood that, for medicine to advance, the study of normal physiology was as important as pathology. At the turn of the twentieth century Pavlov's idea of nervism (see advance 13) was central to the understanding of homeostasis. But before Pavlov

Fig. 15. The chemical messengers (floating above) bind to receptors represented by the three up-facing proteins, which are embedded in the cell membrane depicted by the crosshatched area below.

received his Nobel Prize, William Bayliss and Ernest Starling, two brothers-in-law, discovered an entirely new mechanism of control.

Working in London on their "great afternoon"[18] of January 16, 1902, Bayliss and Starling severed all the nerves to a dog's pancreas and were able to control precisely the secretion of water and bicarbonate coming from the gland's main duct by injecting doses of a "chemical messenger" into the dog's leg vein. Pavlov had already shown that a dog's pancreas begins to secrete water and bicarbonate the instant that gastric acid enters the first few centimeters of the duodenum. So Bayliss and Starling prepared their chemical messenger by taking bits of the mucosal surface that lines the duodenum, boiling them in acid, and obtaining a cell-free acid extract. And with the injection of the extract, the secretory activity of the gland could be regulated by the action of the "chemical messengers" contained within it, even in the absence of any nerves. Moreover, by the inactivation of their extract through the application of the proteolytic enzyme trypsin, which is capable of breaking down proteins containing negatively charged amino acids (lysine and arginine), they knew the messenger to be a protein. They called this protein "secretin" and they fully expected that other such chemical messengers would be discovered. The central paradigm of their newfound mechanism of control was that a chemical messenger could be made in one tissue and released to travel through the bloodstream to find and regulate the activities of a distant target tissue. On that great afternoon, Bayliss and Starling had suddenly understood something about life that had not been known before. They gave the term *hormone*, from the Greek *horman* ("I arouse to activity"), to the class of chemical messengers that act through this paradigm. Within a year, a fellow named Thomas Edkins, working in London along the lines set down by Bayliss and Starling, discovered the second hormone, gastrin, which regulates the stomach's secretion of acid. It was the dawn of endocrinology.

Within a few decades, ACTH (adreocorticotrophic hormone), insulin, PTH (parathyroid hormone), growth hormone, and many other peptide hormones were discovered. Other chemical families of hormones exist, including the steroids and the tyrosine derivatives (thyroid hormones), but the peptides are the largest and most dynamic family. There are about fifty amino acids (of which about twenty commonly occur in proteins (see advance 28). A peptide is a relatively small protein, or a specific sequence of amino acids. The smallest peptide hormone is a tripeptide (only three amino acids), but the number of messages that can be constructed using this language is very, very large. Bayliss and Starling used tryptic digestion to prove that secretin was a peptide. They put trypsin into their extract and it lost all activity to stimulate flow through the pancreatic duct. The two became celebrities, science heroes.

Edkins did the same trypsin test with his extract. Nothing happened; the extract maintained full potency, and Edkins died in disgrace with the scientific world thinking that there was nothing but histamine in the extracts. Many decades later, in the 1960s, when the sequences of these peptides were worked out, it was found that secretin has many lysines and arginines, but gastrin has none. And secretin needs to have all thirty-five of its amino acids in exact sequence for it to work, while gastrin needs only the last four. Biochemistry can be cruel. But not to worry—I have participated in many international conferences where Edkins has been posthumously rehabilitated. PH (posthumous rehabilitation), while not a peptide hormone, is nonetheless an important human tradition.

15

The Germ Theory of Disease

Fig. 16. Louis Pasteur (1822–1895), the great French bacteriologist, worked with rabbits, cows, and people to develop his vaccines.

R emember Leeuwenhoek? (See advance 4.) More than one hundred fifty years after Leeuwenhoek's death, Louis Pasteur (1822–1895) established the germ theory of disease and fought the last and conclusive battle against the idea of spontaneous generation. He developed the science of microbiology, greatly extended Edward-Jenner's theory and practice of vaccination (see advance 16), and created the Institut Pasteur, a worldwide network of medical laboratories, his "temples of the future," so that laboratory science could advance the well-being of humanity.

Yes, the germ theory had been anticipated, even in ancient times:

> When building a house or farm especial care should be taken to place it at the foot of a wooded hill where it is exposed to health-giving winds. Care should be taken where there are swamps in the neighbourhood, because certain tiny creatures which cannot be seen by the eyes breed there. These float through the air and enter the body by the mouth and nose and cause serious disease. (Marcus Varro, circa 46 BCE)

But proving an idea and bringing it into practice is another thing.

Pasteur's life and work created the modern paradigm of the ideal biomedical investigator. He studied challenging problems facing his community, based his approaches upon existing knowledge, used rigorous scientific methods to test his imaginative hypotheses, and courageously brought his research findings to clinical practice. He was passionate in his quest for medical advancement and inspired others to do great work.

Louis Pasteur began his scientific career as a chemist. Using optical instruments to study crystalized tartaric and racemic acids (by-products of wine making), he discovered that the ability of these acids to rotate light both to the left and to the right indicated that they were asymmetric. In 1847, at the age of twenty-six, he formulated a fundamental principle: asymmetric molecules are always organic, the products of living processes. He had initiated the study of stereochemistry, but more important, he set himself on the course of "vitalism," concerning himself only with living processes rather than merely chemical phenomena.

In the 1850s Pasteur was persuaded to learn why wine becomes contaminated with undesirable substances during fermentation. He discovered that fermentation was not simply a reaction among chemicals; it is a process in which specific microorganisms metabolize sugars and other constituents of grape juice, producing alcohols and other products. He studied the alcoholic fermentation of sugar in wine and beer, the production of vinegar, and the souring of milk. This work—seemingly rather narrowly focused on a practical industrial problem—led to revolutionary observations, conclusions, and methods. Pasteur's insistence that the metabolic activities of specific organisms, growing in specific solutions (media), under specific conditions (temperature and oxygen concentration), which result in the production of specific products, initiated the theory and practice of modern microbiology. He discovered the phenomenon of anaerobic life, the world of microorganisms that can live without oxygen. In a series of brilliant experiments he refuted Felix Pouchet's claim to have proven the existence of spontaneous generation, the two-thousand-year-old idea that life could arise spontaneously from organic matter, and in doing so he established the critical importance of meticulous laboratory technique. He showed that the souring of wine and milk could be prevented by destroying the responsible microorganisms in a process of heating the liquid, rapidly cooling it, and then storing it in the cold. This process, known as pasteurization, does not change the flavor or nutritive value of the liquid. Milk is pasteurized by heating to 63°C (145°F) for thirty minutes, rapidly cooling it, and then storing it below 10°C (50°F).

On February 19, 1878, Pasteur spoke before the French Academy of Medicine on his belief that disease, putrefaction, and fermentation were all caused by microorganisms. While such claims had been made even before Leeuwenhoek's discovery of microorganisms, little scientific data were previously available to support the idea. Pasteur had already shown that a disease of silkworms, which was devastating the French silk industry, was caused by a protozoan. He later discovered staphylococcus, streptococcus, and pneumococcus, three bacteria that cause common human diseases. His basic rules of sterilization and asepsis (see advances 31 and 46) revolutionized our concepts of infection and the practice of surgery and obstetrics. His attempts to convince surgeons that contaminated hands and instruments cause infection and that they should wash between cases almost provoked a duel between Pasteur and an eighty-year-old surgeon. Joseph Lister (see advance 31) became a disciple of Pasteur.

Although vaccination was practiced in ancient times (see advance 16), the scientific and laboratory study of acquired immunity, or altered reactivity, began with Pasteur. In a series of dramatic experiments beginning in 1879, Pasteur developed effective vaccines against chicken cholera, anthrax, swine erysipelas, and finally rabies. He was fond of saying that "chance favors the prepared mind," and nowhere was it more appropriate than in his discovery of attenuation. After a vacation in the summer of 1881, he returned to his studies of chicken cholera, caused by what is now called *Pasteurella multocida*. When he inoculated an old culture that had been left in the lab during the summer into chickens, it did not cause disease. Pasteur then inoculated the same chickens with a fresh culture and the chickens remained healthy. Repeated injections failed to cause disease, and Pasteur realized that the aged culture had rendered them immune.[19]

On June 2, 1881, at Pouilly-le-Fort, he was met in the fields by applauding farmers. One month earlier he had immunized six cows, twenty-four sheep, and one goat with a vaccine consisting of cultures of *Bacillus anthracis*, which he obtained from his rival, Robert Koch (see advance 44), and which had weakened through aging. He later injected the animals, and a similar number of unvaccinated control animals, with virulent cultures of *Bacillus anthracis*. When he returned to the fields, the vaccinated animals were all healthy, while all of the controls were dead or ill with anthrax.

From these observations Pasteur constructed the hypothesis that pathogens could be attenuated by exposure to environmental conditions such as high temperature, oxygen, and chemicals. His ensuing work on anthrax and rabies confirmed the hypothesis.[10]

When Pasteur was a boy living in the town of Jura, a rabid dog had bitten several townspeople, all of whom died the gruesome death of rabies victims. Perhaps it was this that turned his attention to rabies. He failed to find the causative agent, a virus that can be seen only with the aid of an electron microscope (see advance 30). Nonetheless, he developed a vaccine by injecting rabies-damaged spinal cord tissue into the brains of rabbits. After successful tests with the rabies vaccine in dogs, a mother brought her nine-year-old son, Joseph Meister, to Pasteur after the lad had been bitten fifteen times by a rabid dog. On July 6, 1885, Pasteur began administering his series of fourteen painful vaccinations of increasingly virulent material. Pasteur was sleepless for weeks, afraid that the boy might not be helped, or that the vaccine itself might kill him. But Joseph Meister stayed well, and, as an adult, he became the gatekeeper at the Pasteur Institute. Fifty years later, when the Nazis occupied France, they came to Meister and demanded entry to Pasteur's crypt. Rather than desecrate Louis' resting place, Joseph killed himself.

16.

Smallpox Vaccination

The Circassians [a Middle Eastern people] perceived that of a thousand persons hardly one was attacked twice by full blown smallpox; that in truth one sees three or four mild cases but never two that are serious and dangerous; that in a word one never truly has that illness twice in life.

—Voltaire, "On Variolation," *Philosophical Letters* (1734)

Protection of individuals and populations from the unparalleled scourge of smallpox represents one of the greatest accomplishments of humankind. It was achieved through scientific discovery, technological innovation, and international cooperation, but it first derived from the folk wisdom of women and the initiative of a courageous mother. In the end, most of the credit has gone to Edward Jenner (1749–1823), a Gloucestershire naturalist and sometime student of surgery, who persisted with his observations and experiments with cowpox. He financed the publication of his results in 1798 after the Royal Society, of which he was a fellow, rejected his findings as "in variance of established knowledge." Still, Jenner had first heard from a local milkmaid that mild cowpox infection, common on the hands of women who milked cows, resulted in effective prevention of smallpox. He also heard that old women at the Ottoman court in Istanbul had been practicing smallpox vaccination for perhaps one hundred years.

Smallpox is caused by a virus (called "variola") of the family

Fig. 17. Lady Mary Wortley Montague (1689–1762) used her lancet to pierce smallpox skin lesions and used a walnut for collection of the pus.

Poxviridae. It is thought to have first appeared in early agricultural communities in northeast Africa about 10,000 BCE. The virus, one-billionth of a meter in size, multiplies only within human cells, causing a disease (variola major) characterized by fever, a disfiguring rash beginning on the face and eyes and finally involving the entire body, as well as back pain, muscle aches, frequently blindness, and often death. The case-fatality rate was generally between 20 and 60 percent, and survivors were usually left disfigured and blind. Among infants and children under five years the fatality rate was 80 to 100 percent. A mild form of smallpox (variola minor) was caused by a less virulent strain of the virus. Long before any attempts at vaccination, it was common knowledge in many parts of the world that a mild case of smallpox would result in protection from subsequent infection. Thus, in many places, children and others were exposed to people with mild cases of smallpox; later materials from smallpox pustules and scabs were administered to the skin of healthy people to confer immunity in a process called variolation. In fact, one hundred years before Jenner, the Chinese were preventing smallpox by oral vaccination, giving healthy people pills made from dried fleas retrieved from cows.

Variolation was a very successful method of protection, producing a minor reaction in the skin at the site, providing complete or partial immunity to smallpox, and markedly reducing case-fatality rates. Jenner himself had been variolated as a boy of eight. Several reports of variolation in travelers to Istanbul appeared in the *Transactions of the Royal Society of London* in 1714 and 1716, but these were ignored or rejected by physicians. The method was brought to England by Lady Mary Wortley Montague, whose husband was ambassador to Istanbul. Lady Montague's face had been disfigured by smallpox, so she ordered her physician in London to inoculate her son in 1718 and then her daughter in 1721. Because of Lady Montague's prominence in society, these led to experiments on prisoners and, after these were successful, to the spread of variolation throughout Europe and to the New World. But it was not without complications. If the material used contained too virulent a strain of virus, individuals would die or become disfigured, which became the source of a new epidemic. If the material was contaminated with tuberculosis, syphilis, or other infectious agents, these other virulent diseases could be transmitted.

Edward Jenner was intrigued by the notion, common among milkmaids, that mild cowpox infection on the hands prevented smallpox, but it was not until 1796 that he tested the idea by inoculating a boy named James Phipps with material from pustules on the hands of a milkmaid named Sarah Nelmes. Six weeks later, and again in several months, Jenner variolated the boy, but the smallpox-derived material produced no local reaction in his skin. He then variolated

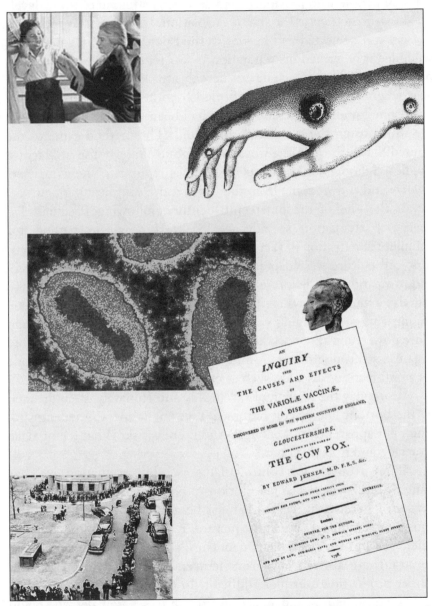

Fig. 18. *Top to bottom*: Child receiving smallpox vaccination; hand with smallpox lesions; smallpox-infected cells; Eygptian mummy scarred by smallpox; poster from a Jenner lecture on cowpox; vaccination lines in England.

thirteen people who previously had cowpox, and none of them developed local reactions at the site of variolation. His reports to the Royal Society were rejected and he was warned not to "promulgate such a wild idea if he valued his reputation."[21] So Jenner published the data himself in 1798, for which he was fiercely attacked, until the following year when his results were confirmed by other physicians. By 1800 about one hundred thousand people had been vaccinated throughout the world using the cowpox pustule fluid. The method came to the United States through the work of Benjamin Waterhouse of Harvard Medical School and the efforts of Thomas Jefferson. The words *vaccination* and *vaccine* come from the Latin root *vaca*, meaning cow.

At the end of the nineteenth century, following Pasteur's discovery of attenuation (see advance 15), work began on methods for complete inactivation of vaccines. By the middle of the twentieth century, cell culture was adapted to growth of viruses (see advance 85), and it was not long before it was realized that passage in cell culture was also a means of attenuation. Today, molecular biology and genetic engineering are changing vaccine development by offering new and more powerful ways of producing inactivated antigens and attenuated organisms through directed mutation. And vaccines are being made for noninfectious diseases, such as cancers.

Also, today the making of vaccines is unfortunately often dictated by the bottom line of corporations. There is money in making vaccines, but apparently not enough. And there is risk. What if you make a vaccine and nobody comes?

"The prospects for control of diseases by vaccination are thus quite bright," writes Stanley Plorkin of Sanofi Pasteur and the University of Pennsylvania, "but it must be admitted that several problems loom large and darken the picture. First, vaccine supply is insufficient. Even in industrialized countries, shortages of vaccines occur because there are too few manufacturers, and regulatory pressures render production ever more difficult. In the event of an emergency, such as an influenza pandemic, it is difficult to see how demand could be satisfied or access provided to developing countries."[22]

Fig. 19. Edward Jenner (1749–1823).

17.

Aspirin

ith all of our advances in knowledge and technology, the most widely used drug in the world has remained virtually unchanged since the fifth century BCE. Aspirin, taken by millions of people each year, with over ten thousand tons used annually in the United States alone, relieves the pain associated with everything from headaches to gout; reduces fever from infection; is a potent anti-inflammatory agent; and as recent studies suggest, may decrease the risk of heart attacks and strokes.

The Greek physician Hippocrates is credited with the earliest known use of a substance now known as salicin, the naturally occurring cousin of acetylsalicylic acid—or aspirin. He employed a powder extracted from willow bark, a tree in the Salicaceae family and high in salicin, to treat pain and reduce fever 2,500 years ago. Today Americans take over twenty billion tablets of aspirin each year, but Native Americans were using the willow tree for a wide range of medical purposes long before the *Mayflower* dropped anchor. It is believed that even primitive man had chewed on the bark and leaves of the willow tree to reduce fever. It is known that throughout the Middle Ages willow leaves and roots were used extensively for the treatment of pain.

Aspirin, which can irritate the lining of the stomach, is a relatively safe drug when taken in recommended dosages. It was developed for commercial purposes in 1898 by Friedrich Bayer and a chemist named Felix Hoffman. Initially attempting to relieve the pain of his arthritic father, Hoffman came upon a salicylate compound—acetylsalicylic acid—which had been synthesized forty-five years earlier by Charles Frederick von Gerhardt. In 1899 the Bayer company named it Aspirin, took out a patent, and soon began selling the drug.

From a scientific point of view, the question of which brand of pure acetylsalicylic acid is "the best" is rather like the one about how many angels can fit on a pinhead. In short, all pure acetylsalicylic acid is the same.

18.

The Initiation of Insulin Therapy

iabetes mellitus means "honey sweet siphon." The name describes a person who passes excessive urine and that urine tastes "wonderfully sweet" owing to its high concentration of sugar. Physicians in the old days would taste urine to diagnose diabetes. In 1674 Sir Thomas Willis distinguished diabetes mellitus from diabetes insipidus by finding that in the latter, while there is excessive urination, the urine lacks a sweet taste.

Among the disorders of the endocrine glands, diabetes mellitus affects the greatest number of people. We currently recognize a number of distinct forms of the disease, but types 1 and 2, resulting from absolute or relative deficiency of the hormone insulin, are the most common.

Insulin is a peptide hormone. A hormone (as described in advance 14) is a chemical messenger that is synthesized and stored in specific cells of a tissue (often called a gland). Under precisely defined conditions, which differ for each hormone, it is released into the bloodstream to travel to some distant target organ(s) where it exercises control over the function of the target tissue.

A peptide is simply a small protein, a relatively short sequence of amino acids that are arranged like beads on a string. There are about fifty amino acids, which can be likened to beads of different shapes and colors; so there are myriad possible peptides, as one might assemble countless different necklaces by varying the sequence in which the beads are strung. A relatively short sequence is called a peptide and longer strings of amino acids are called proteins. The first hormones to be recognized were peptides, and indeed the peptide hormones represent the largest chemical class, but hormones can be of different chemical structures, such as steroids, tyrosine derivatives (thyroid hormones), glycoprotein, and others.

Diabetes is mentioned in the Papyrus Ebers (Egypt, 1500 BCE). In the tradition of Western medicine, it has been known for at least two thousand years that diabetes is a disease in which sufferers experience excessive urination, thirst, fatigue, weight loss, and muscle wasting. In 1889 Oskar Minkowski and Joseph von Mering removed a dog's pancreas and observed that the dog quickly became diabetic. Twenty years earlier, Paul Langerhans had described small clusters of special cells buried within the main substance of the pancreas, and although he did not know their function, these became known as the islets of Langerhans. Shortly after Minkowski and Mering's discovery, Edward Sharpey-Schafer, a great pioneer of early endocrinology who had already made brilliant discoveries relating to adrenalin, found that diabetes resulted from the removal of some active material residing in the islets of Langerhans. He named the material *insuline* after the Latin *insula*, meaning island, but he had no knowledge of its chemical structure. Extracts of pancreatic tissue, including the islets, were prepared and given by mouth in attempts to treat diabetes, but to no avail.

Insulin is a peptide, so, during crude attempts to make extracts of the whole gland, it is rapidly broken down by the protein-digesting enzymes of the main substance of the pancreas. In addition, if such extracts were to contain some undamaged insulin, it would be destroyed before being absorbed into the bloodstream by the proteolytic (protein-digesting) enzymes of the digestive system.

In the summer of 1921, Professor John J. Macleod left his laboratory at the University of Toronto for a fishing vacation in Scotland. He allowed Frederick Grant Banting, a thirty-year-old physician, and his assistant, Charles Herbert Best, a twenty-two-year-old American medical student, to continue working in his absence. During that summer, Banting and Best successfully extracted insulin from the pancreata of dogs and used it to treat diabetic dogs by injecting it into their veins. A key to their success was to prepare the pancreas for insulin extraction by tying off the main pancreatic duct and allowing time for this to cause atrophy of the enzyme-rich portion of the pancreas. In this way, the islets of Langerhans were relatively preserved and insulin could be obtained without contact with the damaging enzymes.

On January 11, 1922, Banting and Best injected insulin into Leonard Thompson, a fourteen-year-old who was dying of diabetes in Toronto General Hospital. Within minutes his blood sugar began to come down, and we now know that all the peptide hormones (secretin, gastrin, cholecystokinin, glucagon, among others) manifest their effects within seconds to minutes after entering the blood. Young Thompson was up and about within a few days. In a short time the Eli Lilly Company was in large-scale production of bovine (cow) and porcine (pig) insulins, saving the lives of countless diabetics, many of whom would otherwise die within days of the onset of diabetes (type 1), as well as others who would have become sick and deteriorated over a matter of months to years (type 2).

Banting shared the 1923 Nobel Prize for Medicine or Physiology with Macleod. Best was only a lab assistant, an undergraduate medical student, so he got no prize. Nonetheless, Banting shared his prize money with Best, and they worked together at the Banting and Best Department of Medical Research in Toronto. There Best went on to discover the vitamin choline and the enzyme histaminase, which breaks down histamine, an important amine found in many tissues of the body. He also became the first to use anticoagulants for the treatment of blood clots.

19.

Florence Nightingale and Modern Nursing

No man, not even a doctor, ever gives any other definition of what a nurse should be than this—"devoted and obedient." This definition would do just as well for a porter. It might even do for a horse. It would not do for a policeman.

—Florence Nightingale (1820–1910)

War, with its gods and heros, is a male tradition and exemplifies much of male culture throughout the ages. In the ancient pantheon, the goddesses Hygeiae and Panaceae signified prevention and treatment of illness. Healing and care for the sick and wounded has been rooted in female tradition, and the development of modern nursing was driven entirely by women in defiance of their exclusion from the world beyond menial work and domesticity.

In the Middle Ages, religious orders, such as the Order of Saint John, and secular orders, like the Third Order of Saint Francis of Assisi, provided some measure of humane caring for the sick; but they did not develop a disciplined nursing profession. By the early Industrial Revolution, much of what passed for nursing was the work of prisoners, pardoned criminals, alcoholics, and aged former prostitutes. Charles Dickens's Sairey Gamp was just such a "nurse," taking

Fig. 20. Florence Nightingale attends to a wounded soldier.

her patients' food, and dinking the families' whiskey, and doing as little else as possible.[23]

By the 1830s the beginnings of organized training for nurses was stirring, primarily within religious communities. Among Catholics, Vincent de Paul founded the Daughters of Charity, and in 1831 Catherine McAuley and Mary Aikenhead founded the Sisters of Mercy and the Irish Sisters of Charity. Within the Protestant community, Theodore and Friederike Fliedner founded the Deaconess Institute in 1836, training women in a small hospital in Kaiserwerth near Dusseldorf. But it was the gross excess of brutality and neglect displayed on all sides in the Crimean War (1854–1856) that inspired a great intellect and spirit to found the modern theory and practice of nursing.

Florence Nightingale (1820–1910) was from an influential upper-class English family. At seventeen she had a religious experience in which she was called "to do something toward lifting the load of suffering from the helpless and miserable." For years she fought her parents' resistance to her desire to work with the sick, until at age thirty-one she was able to go to Kaiserwerth. After three months in Germany, she went to work with the Daughters of Charity in Paris, and in 1853 she became superintendent of the Establishment for Gentlewomen During Illness in London, and then superintendent of nurses at King's College Hospital.

Nightingale had a brilliant mind; she was schooled in history, philosophy, sciences, classical literature, music, and art, and was particularly gifted in mathematics. "To understand God's thoughts," she wrote, "we must study statistics, for these are the measure of His purpose."[24] She traveled extensively; knew French, German, Italian, Greek, and Latin; visited hospitals; and studied their architecture and management wherever she went.

It was with this preparation that Nightingale volunteered to take a group of thirty-eight nurses to the hospital at the English base at Scutari, across the Bosphorus from Istanbul, in November 1854. The British secretary of war accepted her offer over the objections of his generals. His decision was bolstered by public opinion as the British press reported the horrendous conditions under which the death rate for wounded soldiers was 60 percent.

It was seven years before Ignaz Philipp Semmelweis was to write his paper on the cause and prevention of puerperal fever (see advance 31) and nearly a quarter of a century before Louis Pasteur would formally propose the germ theory of disease (see advance 15). Yet Nightingale was already convinced that cleanliness was essential to the hospital environment. She immediately organized her nurses to clean up the place, to provide fresh air, warmth, and quiet, and to pay attention to the diet of the patients. Within six months the mortality rate had fallen to 2 percent. She was concerned with both the physical needs and the social welfare of patients, and instituted, for the first time, sick pay for sick and wounded soldiers. She visited battlefields and made her hospital rounds long into the night, winning the admiration of the press, who referred to her as "The Lady with the Lamp." She returned to England a heroine, but Nightingale suffered from what later came to be known as post-traumatic stress syndrome, so she continued her work in seclusion. Nonetheless, she wrote extensively, and with the support of admirers, many who had served in the Crimean War, she established a school in 1860 for nurses at London's St. Thomas Hospital.

Among the principles that Nightingale espoused were the following: (1) the content of nursing education must be defined by nurses; (2) nurse educators are responsible for the nursing care provided by students and graduates of the nursing program; (3) educators should all be trained nurses themselves; (4) nursing schools should be separate entities, not connected with physicians or hospitals; (5) nurses should be prepared with advanced education and should engage in continuing education throughout their careers; (6) nursing encompasses both sick nursing and health nursing, and involves the environment as well as the patient; and (7) nursing studies must include theory.

Florence Nightingale's conceptual and practical invention of modern nursing as an independent healing profession revolutionized the practice of medicine. Today, we can see her ideas and instructions at work in virtually every hospital ward in the world. But more than that, her exemplary characteristics of intelligence, integrity, compassion, and courage set a standard for healers. We must continue to strive to incorporate all of her wisdom.

However, it would be unfaithful to Nightingale's standard of integrity if we did not emphasize the terrible condition of today's nursing profession. In fact, during my lifetime in medicine, nursing has gone through one crisis after another. Most recently, the managed-care era swept in with the idea that there were far too many nurses in hospitals and clinics. Their numbers were reduced, and so were their salaries. Today, there is uniform agreement that nursing is on the critical list. The numbers of students entering nursing schools has dwindled, and the faculties have shrunk to the extent that they can hardly teach these reduced numbers. The increased numbers of women goining into medicine has not been offset by more men entering nursing. I have spent twenty years teaching classes in nursing school because I believe it is very important, and I have seen this first-hand. The degree of discouragement and poor morale in nursing is painful to witness, and it is a danger to the patient-care environment. This situation has been well documented,[25] and almost any nurse will be sad to discuss it with you.[26]

20.
The Development of Sewage Systems

I counted two and seventy stenches,
All well defined and several stinks!
Ye Nymphs that reign o'er sewers and sinks,
The river Rhine, it is well known,
Doth wash your city of Cologne;
But tell me, Nymphs! what power divine
Shall henceforth wash the river Rhine?

—Samuel Taylor Coleridge (1772–1834)

Here in the United States we turn our rivers and streams
into sewers and dumping-grounds, we pollute the air, we
destroy forests, and exterminate fishes, birds, and mam-
mals—not to speak of vulgarizing charming landscapes
with hideous advertisements.

—Theodore Roosevelt (1858–1919)

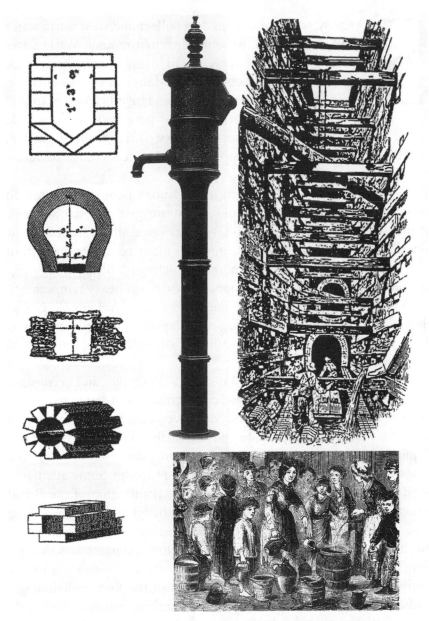

Fig. 21. The Broad Street Pump (*center*), the Fleet Street Sewer (*right*), the water supply in Fryingpan Alley, where water was available for twenty minutes each day, and several early designs for sewers (*left*).

Sewage systems are designed to collect and treat waste water before it is released back into the environment. Waste water includes the used-water effluent from homes, businesses, and major industrial operations. Many sewage systems also handle the flow of storm drainage. The efficient working of these systems is essential to the health of people living in sizable towns and cities. Nonetheless, they have often been neglected, resulting in disastrous epidemics of disease, necessitating the rediscovery of basic principles.

Between 3000 and 1500 BCE, the Minoan people of Crete laid elaborate systems of sewage disposal and drainage that resemble those of today. In fact, archaeologists have discovered underground channels that remained virtually unchanged for many centuries, apart from basic extensions added to incorporate structures built over the original ones. Some vestiges of the pipes still carry off heavy rain water to this day. In many ancient cities, like Rome, city development can be closely correlated to water management systems, including sewage. The ancient Roman Cloaca (common channel for urine and feces) still functions.

Modern metropolitan sewage systems are complex and in constant evolution, as pressures from population growth and industrial development challenge their capacities. Moreover, catastrophies of the past serve to stimulate the push for advances in methodology. But the requirement for keeping waste water from contaminating drinking water remains the fundamental principal. Dr. John Snow attributed the 1854 London cholera recurrence to contamination of the Broad Street Public Well by local privy vaults. He had published *The Mode of Communication of Cholera* in 1849, but most scientists still subscribed to the "miasma" (bad air) theory. The well below the pump was twenty-eight feet deep, and at twenty-two feet, there was a sewer within a few yards of the well. The water stank. In one of the first epidemiologic studies, Snow traced almost every one of the three-hundred-some-odd cases of cholera that developed between August 31 and September 2 and found that they all had drunk from the Broad Street well. That this required a brilliant observation, which came nearly a quarter century before Pasteur emphasized the germ theory of disease (see advance 15) and initiated the science of epidemiology,[27] emphasizes how much had been lost to our culture concerning basic knowledge of sanitation.

Snow's discovery prompted local officials in London, as well as in other European cities, to order that human waste be discharged into the storm drainage runoff. The result was protection of some public wells, but at the cost of the gross pollution of the Thames and other rivers, which produced more waterborne epidemics, requiring the development of methods for the treatment of sanitary waste.

An early method of waste treatment was simply to spread raw sewage on the land, allowing the filtering and absorptive capacity of the soil to return purified water to the aquifer. This also provided organic fertilizer, so this method enjoyed more than a century of general use. The Industrial Revolution, which concentrated people and industry, also intensified the problem of water pollution and prompted chemical treatment methods and the separation of storm and waste-water sewers. More recently, recycling by treating sewage water to a level allowing its use for irrigation, industrial cooling, and other industrial processes has been employed, especially in arid regions. But the sheer volume of sewage demanded newer methods.

Another method used large concrete tanks filled with fist-size stones. Waste water would be allowed to settle into the bed of stones for several hours where organic matter clung to the interstices of the massive bacterial growths that had developed on the stones. The waste water would be drained through a series of such tanks in the process. This led to trickling or sprinkling filters in which waste water trickled from one rock tank to the next.

With the understanding that oxygenation of organic wastes is essential to the purification process, British workers developed the "activated-sludge process" in which waste water is oxygenated in large tanks through the introduction of compressed air. The resulting mixture, teeming with bacteria, protozoa, and chemicals, is known as activated sludge. This method is thought to provide a more controlled and intense oxygenation (exposure to oxygen), allowing a more efficient and higher level of treatment. It has spawned a "biosolids" or sludge industry in which this material is spread on agricultural and other plots of land. But neither the producers nor the consumers know exactly what the material contains.

The provision of quality drinking water has been always and everywhere a highly political issue and closely tied to the problem of sewage. Whether it is the provision of water in the development of Los Angeles, or the watershed in the Catskill Mountains for New York City, or the sanitation of the Jordan river or that of the Ganges, most people, polluters, politicians, entrepreneurs, governments, and corporations, all use or ignore this issue concerning basic hygiene. In this regard, it should be understood that corporations tax the public by polluting, that many parts of the world are nearing or have long passed critical requirements for improvement of water supplies, and that a growing number of people are convinced that their health depends upon paying good money for bottled water from "natural" sources. This would undoubtedly have disappointed John Snow, who spent his life working to improve public health and public water supplies.

21.

Public Works and Health

What vast additions to the conveniences and comforts of living might mankind have acquired, if the money spent in wars had been employed in works of public utility; what an extension of agriculture even to the tops of our mountains; what rivers rendered navigable, or joined by canals; what bridges, aqueducts, new roads, and other public works, edifices, and improvements . . . might have been obtained by spending those millions in doing good, which in the last war have been spent in doing mischief.

—Benjamin Franklin (1706–1790)

Maintaining the health of individuals in communities, which have grown beyond a few thousand people, is not possible in the absence of public works. Public works are government-sponsored facilities built for public utility, convenience, or enjoyment. They are the bedrock of civilization, among the oldest and most defining institutions of human culture. In addition to roads, highways, bridges, canals, inland waterways, dams, harbors, lighthouses, and tunnels, essential for industry and commerce, there are public schools and parks. Public works include water-supply systems, sewage systems, refuse-disposal systems, and other public activities that are essential for the maintenance of public health.

In some places public works have become so familiar that they are

99

taken for granted. The nearly century-old system that has provided New York City with high-quality free water and New York's mammoth sewage and water treatment facilities are good examples. But the importance of public works becomes even more clear in the face of disasters like earthquakes, floods, or wars, when large groups of people are forced to live without public facilities. Disease and corruption are immediate consequences. The history of Europe is replete with examples of how poor sanitary arrangements result in deadly epidemics of cholera, plague, and other diseases.

It is important to be aware that public works are frequently under attack by those who would give essential public facilities to business interests for profitable exploitation. This fundamental distortion of democratic tradition poses great danger to the public health. Although budget reductions and official mismanagement of public works initially hurts poor members of the community, in short order everyone is effected. It is the responsibility of every citizen to protect public works such as water supply, sewage, and waste treatment and disposal systems, which are certainly among the greatest advances for the protection of health. The outbreak of mosquito-borne viral encephalitis that has threatened New York City residents since the summer and fall of 1999 is a good example of the consequences of neglect of public health institutions. After more than a decade of budget reductions for public health, New York City was left with no program to monitor and control mosquito populations. The importance of such programs has been well known for over a century, and public health officials have agreed that the outbreak was a disaster waiting to happen. The city's response to the outbreak was wholesale spraying of pesticides, a desperate attempt to make up for years of neglect, and it represented an additional potential threat to everyone in the population.

The fact that many politicians at every level of government are rising to denounce public works, suggesting that we close public hospitals and sell or give away public water supplies, should cause us to consider the great medical advances that many of these represent.

Paul Ehrlich's "Magic Bullets"

> **It is unthinkable for a Frenchman to arrive at middle age without having syphilis and the Cross of the Legion of Honor.**
>
> —André Gide (1869–1951)

In the year 2001, members of the Nobel Foundation were looking back at a century of choices of people who were deemed to have brought the greatest benefit to humankind. Rolf Luft, a senior among participants in the Nobel community, a brilliant physician-scientist (see advance 77), and a past chair of the Nobel Committee for Medicine or Physiology, rates Paul Ehrlich (1854–1915) at the very top in terms of his "creativity." Indeed, considering Paul Ehrlich poses a unique dilemma because he made so many conceptual and practical contributions of great significance that focusing on one is difficult. But that underlies the greatness of his genius. Here is the short list of his credits:

- Coined the term and concept of "magic bullets"
- Conceptualized the cell membrane receptor
- Produced the first cure for syphilis
- Laid the foundation for the science of immunology
- Founded chemotherapy
- Founded hematology
- Cured diphtheria with antitoxins

He was the quintessential idea man, a theorist rather than a great experimentalist.

In the first decades of the twentieth century, there were few effective chemical treatments for disease. Of course, the painkilling drugs morphia and aspirin were in use, ether and chloroform provided chemotherapeutic anesthesia, digitalis "strengthened" the heart, amyl nitrate dilated the coronary arteries in relief of painful angina, colchicum was used with success for gout, iron was considered a popular tonic, senna and other plant-derived purgatives were applied, quinine was taken for the treatment of malaria, and mercury was given for syphilis and ringworm. Most of these derived from traditional herbal remedies. Physicians had an emerging understanding of health and disease, but little of therapeutic value. Nonetheless, the germ theory of disease, which provided specific "enemies" to aim at, and the emergence of immunization (in which antibodies produced in response to specific microbes were toxic only to those microbes while harmless to the patient), would soon produce the theoretical stimulus for the current movement in therapeutics toward "designer" drugs. Ehrlich, with his background in bacteriology and immunology, and his close relationship with the German chemical industry, jumpstarted this revolution.

The center of Paul Ehrlich's contributions is the idea of specificity: the notion that chemical structure allows unique biological activity. He was impressed with the specificity of antigen-antibody reactions and the selectivity of tissue staining with methylene blue and other aniline dyes. He believed that specificity of action was related to chemical structure, that a molecule became fixed, or attached, to a cell membrane, or microbe at a special receptor site, the way a key fits into a lock. He made conceptual drawings of his putative receptor sites on cell membranes, and searched for "magic bullets," chemicals that would find targets on pathogenic microbes and kill them without harming the host. Because arsenical compounds, while very toxic, were known to have mild action against *Treponema pallidum* (the organism that causes syphilis), Ehrlich synthesized and tested over six hundred arsenicals for anti-*Treponema* activity. Preparation #606 had strong activity. It was named Salvarsan, modified to Neo-Salvarsan, renamed neoarsphenamine, and although quite toxic and requiring many painful injections, it was by 1910 the first effective treatment for syphilis.

Neo-Salvarsan was hardly a magic bullet, but it was a start. In 1927 workers at I. G. Farbenindustrie, the powerful chemical company that had acquired Bayer and Hoechst, found that one of their azo dyes used for coloring fabric, Prontosil red, cured mice infected with streptococci. (I. G. Farben also became the maker of Zyklon B, the poison gas used in the Nazi extermination camps, and a major exploiter of slave labor during World War II.)[28] Studies at the Pasteur Institute confirmed the finding and determined that the sulfanilamide portion of Prontosil, while not lethal to streptococci, prevented the organisms from multiplying. This "bacteriostatic" action allowed the patient's immune system to dispose of the bacteria. Thus, we had the first of the sulfur drugs, the first "wonder drug," and something like a magic bullet that could find its targeted bacteria and kill them without harming normal cells. Nearly as wonderful as its effects, it was readily available because Prontosil, which was synthesized in 1907, could not be patented.

23.

Eradication of Smallpox by the World Health Organization

Once smallpox scourged the earth, spreading a thousand years BCE through Egyptian merchants, from North Africa through India, and finally, with trade and conquest, to all but the most remote human populations. The faces of Egyptian mummies dating back 1600 years BCE were scarred by smallpox, and the first recorded epidemic occurred in 1350 BCE during the Egyptian-Hittite war. The Hittite king and his son and heir were among the victims, and their civilization was the first of many to fall as a result of smallpox. In 180 CE a smallpox epidemic that killed nearly seven million people initiated the decline of the Roman empire. The Spanish and Portuguese brought smallpox to the New World where it ravaged the Aztecs and the Incas. Within one hundred years of the Spanish arrival in Mexico in 1518, the native population was reduced from twenty-five million to less than two million. All over the world, colonialist expansion had similar results. In 1763 Sir Jeffrey Amherst, commander of British forces in North America, invented biological warfare by suggesting that material obtained from smallpox scabs and pustules be ground into blankets to be given to tribes of Indians. And this practice was carried out on the Plains Indians by the American government in the nineteenth century. By the twentieth century smallpox had spread to about half a billion people, more than in every war and all other epidemics combined.

After the demonstrated success of Jennerian vaccination (see advance 16), the practice spread to many parts of the world, but some time during the nineteenth century the cowpox virus was replaced by the related vaccinia virus as the starting material for the production of vaccine in cows. A vaccine consists of live, attenuated, or killed microorganisms (sometimes just a fragment of the organism's specific protein), which, when given to someone by mouth, injection, or inhabitation, causes the production by lymphocytes of specific antibody proteins that can neutralize or kill the offending organism or closely related species. In 1926 the Smallpox and Vaccination Commission of the Health Organization of the League of Nations began to study the production, testing, standardization, storage, and delivery of smallpox vaccine. By the mid-1900s smallpox was under control in Europe and North America, but it was rampant in many other parts of the world. Human antivaccinia immunoglobulin, prepared from the blood plasma of recently vaccinated people, became available in the early 1950s. This material could be administered at the time of vaccination to prevent illness in people who fail to produce sufficient amounts of their own antivaccinia antibodies. So the availability of standardized vaccines, the production of human antivaccinia immunoglobulin, and the success of smallpox eradication in Europe and North America set the stage for a global effort to eradicate smallpox. Because the variola virus can live only in human cells, eradication from all human populations could result in the extinction of the virus.

In 1958 Professor Viktor Zhdanov of the USSR, a state that shared long borders with regions in which smallpox was still endemic, proposed a global program to eradicate smallpox. At that time, the WHO regional program to eradicate smallpox in the Americas was the only cooperative international effort against the disease. In May

1959 the Twelfth World Health Assembly undertook global eradication. The effort in the Americas, begun in 1950, was languishing, and it was not until 1966 that the Nineteenth World Health Assembly allocated WHO funds for the global effort.

Even then there was great reluctance to proceed, and many believed that eradication of smallpox was unachievable. UNICEF refused to support the program. Widespread skepticism was fed by the immense logistical problems, especially in highly endemic areas with scant reporting and public health capability, and by the failure of attempts to eradicate malaria. But the availability of highly effective freeze-dried vaccine and the fact that smallpox virus can live only in human cells provided the opportunity for success in eradicating it.

Between 1967 and 1980, with the combination of mass vaccination campaigns and the development of a surveillance system for detection and investigation of all cases of smallpox throughout the world, planners achieved success. The USSR contributed over fourteen million doses of vaccine and the United States contributed 190 million doses. The last case of smallpox was recorded on October 26, 1977, in Merca, Somalia.

At this time there are only two known remaining laboratory sources of the smallpox virus: one in the Moscow Research Institute in Russia and the other at the Centers for Disease Control in Atlanta, Georgia. Although the genetic sequence of the virus is known, a debate now rages over whether these last sources of smallpox virus should be destroyed or saved for study. In fact, with the suspension of vaccination programs after the eradication of the disease, world populations are becoming more vulnerable to possible new outbreaks of smallpox, and fear has arisen that terrorists or governments might be saving cultures of smallpox for biological attacks.[29] The date for destruction of known smallpox virus was put off from June 1999 to June 2002, and the debate continues over the first planned extinction of a species.

24.

Gregor Mendel's Discovery of Genetics

Fig. 22. Gregor Mendel (1822–1884) with pea pods and hybrid pea plants in which the gene for larger size is dominant.

The theory of evolution by natural selection was delayed for nearly twenty years because Charles Darwin anticipated the furious reaction that a challenge to the biblical story of creation would provoke. It was first presented on July 1, 1858, in a joint paper given by Darwin and Alfred Russel Wallace to the Linnean Society in London. But it would take another six months and the publication of Darwin's *The Origin of Species* before the shock of these ideas would reverberate around the world. The theory holds that evolution results from variation and selection, plus inheritance. But while Darwin and Wallace discussed selection, they never offered a satisfactory explanation for variation or inheritance.

We owe the origins of an understanding of variation and inheritance to an obscure Moravian, born Johann Mendel in the hamlet of Heinzendorf in 1822, who took the name Gregor in 1843 when he became a novice monk. Mendel initiated the science of genetics and charted a course for genes, DNA, the genome project, gene therapy, and beyond.[30] So advanced was Mendel in his thinking that he was well into his great work before Darwin published *Origin*, and it was not until the first month of the twentieth century, some twenty years after he completed his studies, that it was rediscovered and understood.

It appears that Mendel, the son of a farmer, entered the church to obtain an education and pursue a studious life, but little is known of his full motivation. In becoming a passionate and well-informed scientist, he began studying heredity by crossbreeding mice and observing the inheritance of their characteristics. But if science itself was regarded with suspicion by the church, these experiments probably pushed too far, and he soon selected the common garden pea for quiet research in the Augustinian monastery garden at Brunn. His focus was on the hybridization of a few varieties of peas that were easily fertilized and had clearly distinguishable pairs of characteristics. He studied pairs of color using peas with purple flowers and peas with white flowers, peas that are yellow and peas that are green, and surface type using peas with wrinkled skin and smooth peas. He limited his studies to seven such pairs of characteristics, crossing "wrinkled" mother with "smooth" father, and "smooth" mother with "wrinkled" father, and crossing their offspring with each other and with each type of parent, and so on, and on. His work was characterized by meticulous attention to detail in the characterization and fertilization of his plants, repetition, and rigorous statistical methodology. He took his hybrid peas through multiple generations to follow the influences of heredity.

In one of his studies, Mendel used plants that always produced green seeds and cross-fertilized them with plants that always produced yellow seeds. He did not get plants that produced seeds of some blended color. In fact, all the plants produced yellow seeds. But when those yellow seeds, which were the product of cross-fertilization, were grown into plants that were then allowed to fertilize themselves naturally, the "grandchildren" of the original pure strains of yellow and green peas produced pods in which five or six peas were yellow and two or three were green. The actual numbers that he reported were these: of 8,023 peas, 6,022 (75 percent) were yellow and 2,001 (25 percent) were green. Now Mendel planted these peas and found that the green peas produced plants with pods containing only green peas, but the yellow peas produced plants in which one-third had only yellow peas, while two-thirds had both yellow and green peas in the ratio of 1:3.

Mendel's explanation for these data were that yellowness and greenness are passed through the generations as characters (now called genes), which are inherited in a double dose in each pea. One dose comes from the male parent and one dose comes from the female parent, and one form of the character (gene) dominates over the other recessive form. In 1865 he wrote, "Those characteristics that are transmitted entire, or almost unchanged by hybridization, and therefore constitute the characters of the hybrid, are termed dominant, and those that become latent in the process, recessive."[30]

We can call the gene for yellow A and the gene for green a. In current terminology, these two variations of the gene are called alleles. A pure-bred yellow pea would have inherited identical alleles from each parent and its genotype (or genetic "blueprint") for seed color would be represented as AA. A pure-bred green pea would have an aa genotype. When an AA plant fertilizes an aa plant, the seeds produced inherit one allele from each parent and thus are Aa and all are yellow because A is the dominant allele, the only one expressed.

Since each new plant inherits one allele from each parent, hybrid Aa crosses result in 50 percent Aa, 25 percent aa, and 25 percent AA seeds. And the phenotype, or the observable expression of the genome, is that 75 percent of the peas will be yellow and 25 percent green. The same statistical analysis predicts the proportion of each type of pea in subsequent generations.

While complex biological forms have few clear-cut systems of dominant and recessive genes, Mendel's work is central to our understanding of evolution and the foundation of modern genetics. In clinical medicine we define certain characteristics as having Mendelian inheritance. Mendelian diseases result from a single mutant gene that has a large effect on phenotype, and they are inherited in patterns similar to those described by Mendel for discrete characteristics in garden peas. In 1866 when Mendel published his results and conclusions, the scientific communities of the world were unimpressed and took little notice. Only the church reacted, regarding him as a Darwinian, so Mendel left science and retreated into his monastic work. He had done it all in seven years. He died in 1884. In 1999, with issues of genetic engineering at hand, the state of Kansas banned the teaching of evolution, and the argument over the monk's science has since intensified in America.[32]

Friedrich Miescher
Discovers DNA in Pus

The great pathologist and liberal politician Rudolf Virchow (1821–1902) was among the first medical scientists to focus attention on the importance of cells. He had been forced from Berlin to Wuzburg because of his support for the revolution of 1848. And later he cofounded the Progressive Party and was elected to the Prussian House of Representatives in 1860, where he had violent clashes with Otto von Bismarck. In science, he also generated controversy. Virchow contended that "pus cells" did not form de novo (from nothing) at the site of inflammation but were simply white blood cells recruited from the bloodstream. But while Virchow focused attention on cells, it was Ernst Haeckel, in 1866, who suggested that the nucleus of the cell held the information for heredity.

In 1868 a young physician named Friedrich Miescher went to work in the laboratory of physiological chemistry at Tubingen because he was convinced that great insights could be achieved by studying the biochemistry of simple human cells. He reasoned that the easiest way to obtain great numbers of cells for study was from the pus-soaked bandages to be found on every surgical wound. In the course of washing white blood cells from the bandages, he found that weak alkaline solutions caused cell nuclei to swell and burst, resulting in the appearance of a new type of chemical in the solution. Within a year Miescher discovered that this new chemical could be found in yeast as well as in the nuclei of cells from many tissues. He called it nuclein and noted that in addition to carbon, hydrogen, oxygen, and nitrogen, it contained phosphorus.

The Franco-Prussian War and the machinations of the egotistical Professor Felix Hoppe-Seyler delayed the publication of Miescher's work, which appeared in 1871 in Hoppe-Seyler's journal called (what else?) *Hoppe-Seyler's Journal of Medical Chemistry*. Nevertheless, Miescher had discovered nucleic acid.

Although a zoologist named Oscar Hertwig declared in 1884 that nuclein was responsible for the transmission of hereditary, an understanding of the complex structure of the nucleic acids—ribose nucleic acid (RNA) and deoxyribose nucleic acid (DNA)—and a general appreciation of their importance would take many decades more. In 1909 a group at the Rockefeller Institute for Medical Research (now Rockefeller University), led by Phoebus Levene, a Jew who had fled Russia in 1891 to escape persecution, discovered that ribose was the sugar contained in nucleic acids. In the 1920s it was found that one form of nucleic acid has ribose that lacks one oxygen atom (hence deoxyribose). So, by the 1920s it was known that sperm cells were essentially bags of DNA, that DNA was constructed of chains of deoxyribose connected by phosphate groups, and that each sugar molecule held one of four purine or pyrimidine bases (purines = adenine [A] and guanine [G]; pyrimidines = cytosine [C] and thymine [T]) to create a sequence of repeating units. But a message that runs GCAT GCAT seemed too simple to be the blueprint for constructing all the substances of life. And it was left to Oswald Theodore Avery to uncover the function of DNA.

26.

Oswald Theodore Avery Uncovers the Function of DNA

Everyone knows that DNA is of paramount importance. In terms of life on earth, DNA is the most essential substance. DNA defines you, me, and all the animals, plants, bacteria, you name it: if it's alive on Earth, DNA is in control. The imagination and the work of nearly every biomedical researcher is focused on DNA.

Although nucleic acid was discovered in 1871 (see advance 25), DNA was a bit player until 1944 when Oswald Theodore Avery and his young associates, Colin Macleod and Maclyn McCarty, proved that DNA is the hereditary material, the substance of genes. Avery was then sixty-seven years old and had already retired as an emeritus member of the Rockefeller Institute. He would retire to Nashville in 1949 and live in relative obscurity until his death in 1955. Sir Peter Medawar (see advance 50) called Avery's discovery "The most interesting and portentous biological experiment of the twentieth century."[33] Joshua Lederberg, who won the Nobel Prize in 1958 for his contributions to genetics, called it "the historical platform of modern DNA research. . . . [It] betokened the molecular revolution in genetics and biomedical science generally."[34]

In 1866 it was already appreciated that a sperm cell is little more than a motile nucleus. Moreover, the German zoologist and philosopher Ernst Haeckel met Charles Darwin to suggest that the material that transmits hereditary information was in the nuclei of cells. By the 1880s the penetration of an egg by a sperm and the fusion of their nuclei to form a new nucleus had been observed under a microscope and described. And not long after the discovery of DNA it became clear that sperm contains a bit of simple protein and a lot of DNA. Indeed, the idea that the material responsible for heredity was DNA had been offered by several individuals. But there was no proof, and there was a more popular candidate for this distinction. DNA seemed too simple to contain all the complex information required. Proteins, on the other hand, were known to be of immense importance; they are structurally complex, and they transmit complex information, for example, as hormones. Proteins were the leading candidate for the hereditary material, and the idea was that the nucleus must contain a small amount of every protein to serve as a template that could be copied. Though only a few very simple proteins could be found in nuclei, the excitement surrounding the rapid development of protein chemistry kept them in the spotlight until Avery and his pneumococci shined the big light on DNA.

In the 1920s bacteriologist Fred Griffith of the Ministry of Health in London was trying to understand why some pneumococci cause disease while others are harmless. It had been observed that bacteria frequently grow in colonies that appear either rough (R) or smooth (S). The S colonies are coated with a sugar and are frequently virulent, while the R colonies are generally harmless. Griffith found that if you kill an S colony of pneumococci, they, of course, become harmless, but if you inject mice with a mixture of some dead S pneumocci and a live, but harmless, R colony, the mixture will be lethal. Moreover, the bacteria recovered from the injected mice are S-type pneumocci, which breed true in subsequent generations. How had the harmless R bacteria been transformed into lethal S organisms? What was the transforming factor? Since pneumoccal infection was then the leading cause of death, this was an important issue for clinical medicine. It also went to the fundamentals of genetics. Furthermore, since bacteria breed so rapidly, requiring only minutes for each generation, they became tools for studying heredity.

From 1935 to 1944 Avery's group showed that the transforming material in the S colonies was not a protein, a fat, or the sugar coating that gave S colonies their smooth appearance. And after removing all proteins, sugars, and fats, the S colony's transforming material contained only nucleic acid. Their chemical analysis revealed that the nucleic acid was DNA rather than RNA. And so "the platform for modern DNA research" had been constructed, but the Nobel committee failed to recognize its importance, and Oswald Avery faded away.

27.

The Structure of DNA Is Elucidated by James Watson, Francis Crick, Maurice Wilkins, and Rosalind Franklin

Take young researchers, put them together in virtual seclusion, give them an unprecedented degree of freedom and turn up the pressure by fostering competitiveness.

—James D. Watson on his formula for breakthroughs in research

In a nonscientific survey among medical school faculty physicians, I have found that the elucidation of the structure of DNA, along with the genetic code, is most often selected as the number one medical advance. This is often expressed in terms of the technical project that flowed from these and related advances: the human genome project. These discoveries are often regarded as "giving us the secret of life" or "giving us the power to play God." This assessment generally forced us to reevaluate our criteria for greatness as well as the impossible task of ranking medical advances. Nonetheless, we are ranking great advances, not important

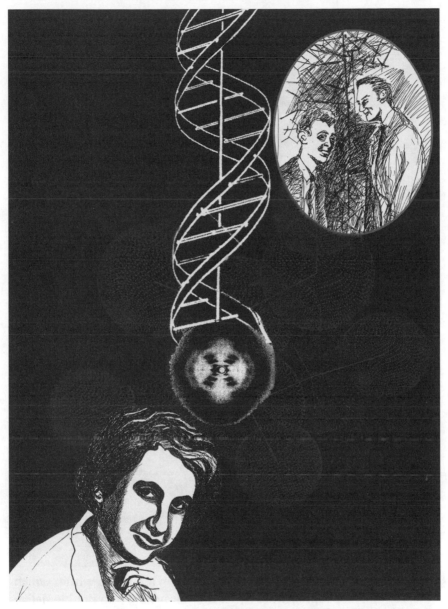

Fig. 23. DNA researcher Rosalind Franklin (1920–1958) (*below*) with her x–ray diffraction photograph 51, and with a conceptual drawing of the double helix extending upward from it; An image of James Watson and Francis Crick (amino acid alanine is in the background) is above right.

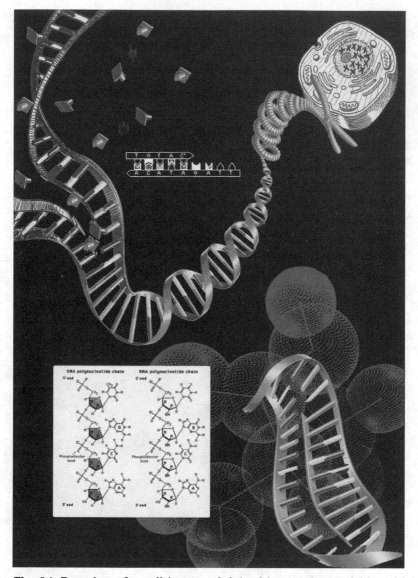

Fig. 24. Drawing of a cell (*upper right*) with a nucleus within which chromosomes represented by Xs are unraveling into double helix DNA strands, which further split into single strands that are shown replicating. Chart below showing the purine and pyrimidine bases, also a cartoon (*upper center*) showing the base pairing scheme in which adenine on one chain is paired to thymine on a complimentary chain and cytosine is paired with guanine.

discoveries. An advance can include the way the field of medicine moves into the future while a discovery is the understanding of a part of nature for the first time. We do not yet know what societies will make of the power to manipulate genetic structures.

No matter what the ranking, these discoveries have a special cache, a mystique, a *je ne sais quoi*, not unlike Einstein's $E = mc^2$. They represent such fundamental insights that they must be viewed on a universal scale: whereas one clarifies the physical universe by describing the relationship between matter and energy, the other reveals the biological universe by unifying all life-forms. Both play out at the core of things—the nuclei of atoms and cells. Both display an elegant simplicity. Both bind the most remote past with the most distant future.

Today an understanding of the basic structure-functional relationships of DNA is required for an informed life. Today the structure of nucleic acids (DNA and RNA) seems simple. But in the past an integrated understanding of biologic, chemical, and physical sciences— from genetics to hydrogen bond steriochemistry—was necessary before the eureka moment would arrive at the Cavendish Laboratory of Francis Crick and James Watson in 1953. From that moment on, macromolecular structure would be the heart of biology. Only a few years earlier the revolutionary chemist and author of *The Nature of the Chemical Bond*, Linus Pauling, had discovered (along with Vernon Ingram) that all the manifestations of sickle cell anemia were caused by the substitution of one amino acid for another (valine for glutamic acid) in the sequence of hemoglobin. He had also made other contributions crucial to the discovery of the nucleic acid structure, but he was effectively hampered in his scientific work by the US government.[35] They harassed him and denied him passports because of his positions against war and nuclear weapons. At the same time and for the same reasons, they also denied an entrance visa to Dorothy Crowfoot Hodgkin, the British scientist who had discovered the structures of insulin, vitamin B-12, and other key molecules.[36]

But others, especially Watson, knew the essence of Pauling's thinking: his discovery of alpha helical protein structures, his ideas about hydrogen bonding within macromolecules, his use of molecular model making, and his insistence on using intuition as a scientific aid. All of these were central to solving the DNA problem. Watson was a biologist who had worked with bacteriophages and tobacco mosaic virus (TMV). Maurice Wilkins was a physicist who, during World War II, had worked on the atomic bomb project, but after its use, like many others, he turned his attention to biologic questions. Crick also had a background in physics and did early work on x-ray diffraction of proteins, discovering the alpha helical structure at about the same time as Pauling. Franklin worked in Wilkins's laboratory and made crucial contributions to the x-ray diffraction work, which in revealing the shapes and angles of the DNA molecule, along with the critical chemical composition data of Erwin Chargaff, was essential to the construction of the correct three-dimensional model of DNA. Chargaff had discovered that while the sequence of bases along the polynucleotide chain was complex and the base composition of varying DNAs differed, the number of adenine and thymine groups were always equal, and so were the number guanine and cytosine groups.

In his Nobel lecture, Maurice Wilkins stated:

> The key to DNA molecular structure was the discovery by Watson and Crick that, if the bases in DNA were joined in pairs by hydrogen-bonding the overall dimensions of the pairs of adenine and thymine and guanine and cytosine were identical. This means that a DNA molecule containing these pairs could be highly regular in spite of the sequence of bases being irregular. Watson and Crick proposed that the DNA molecule consisted of two polynucleotide chains joined together by base pairs. Watson and Crick built a two chain model of this kind, the chains being helical and the main dimensions being as indicated by the X-ray data. In the model, one polynucleotide chain is twisted round the other and the sequence of atoms in one chain runs in opposite direction to that in the other. As a result, one chain is identical with the other if turned upside down, and every nucleotide in the molecule has identical structure and environment. The only irregularities are in the base sequences. The sequence in one chain can vary without restriction, but base-pairing requires that adenine in one chain be linked with thymine in the

other, and similarly guanine to cytosine. The sequence in one chain is, therefore, determined by the sequence in the other, and is said to be complementary to it.[37]

Thus, genetic information is written in a four-letter language as a sequence of bases along a polynucleotide chain. This model allowed understanding of how genetic material makes exact copies of itself, how growth proceeds in an orderly fashion, how life originated, and how favorable characteristics are preserved through natural selection.[38] It is the essence of all existence. Since 1980 the US Patent and Trademark Office (PTO) has awarded patents on the cellular structure of living organisms. Recently, the subjects of these patents have moved to human cells. For many reasons, the issue of patenting genes has spawned controversy.

Advances like these have profound social and political implications. They influence our existence in ways we can't begin to comprehend, unless we understand some basic information. It is dangerous to separate the science from the social and political implications of such great advances. But that is what often happens. Consider poor Alfred Nobel, whose invention of dynamite made him one of the wealthiest men of his time. But its weaponization gave him recurrent nightmares with no end to his grief. He became a tormented soul. He then became close to Bertha von Suttner, his former secretary, who had become a well-known pacifist and author of the book *Lay Down Your Arms*. It was their concerns that prompted Nobel to write his 1895 will establishing a fund, "the interest on which shall be annually distributed in the form of prizes to those who, during the preceding year, shall have conferred the greatest benefit on mankind."[39] We should not ignore the implications of invention.[40] Many leading scientists and scientific organizations are on record regarding the requirement to disclose and discuss the social and political implications of advances. What will become of our DNA sequence information?

28.

The Genetic Code

To see a world in a grain of sand
And heaven in a wild flower
Hold infinity in the palm of your hand
And eternity in an hour.

—William Blake (1757–1827)

e now know that the function of DNA is to replicate living organisms and direct the synthesis of proteins. Proteins are sequences of amino acids (think of amino acids as beads of distinct shapes), each connected to the next by a peptide bond (the string between the beads). The sequence of the beads is the crucial element that determines the character of the molecule; its shape; its charge; where in life it may go; with which molecules it may interact, bind, stimulate, inhibit, do business; its function. The amino acid sequence of any protein is determined by the sequence of bases in some region of a particular DNA (nucleic acid) molecule (a gene). There are about twenty different amino acids that are commonly found in proteins, and four bases occur in nucleic acid. As Francis Crick stated in his 1962 Nobel lecture, *On the Genetic Code*, "It is one of the more striking generalizations of biochemistry . . . that the twenty amino acids and the four bases, are, with minor reservations, the same throughout Nature."[41]

How does a sequence of four things determine the sequences of

122

Fig. 25. The twenty-three pairs of human chromosomes shown in the box along with a rendering of human chromosomes shown over a strip of gene sequencing gel.

twenty or more things? The genetic code describes the way. Watson and Crick quickly calculated that with groups of three bases (triplets) there could be sixty-four possible sets, while pairs would yield only sixteen different sets. Since, as mentioned above, proteins of all life-forms contain at least twenty amino acids, they figured that triplets of bases could code for one amino acid. They called these sets of triplets "codons." In the short hand of this scheme, the bases are represented by their first letters (A = adenine, T = thymine, C = cytosine, G = guanine), and the amino acids have standard abbreviations as shown below:

Amino Acids

Name	Abbr.	Linear structure formula
Alanine	ala a	CH3-CH(NH2)-COOH
Arginine	arg r	HN=C(NH2)-NH-(CH2)3-CH(NH2)-COOH
Asparagine	asn n	H2N-CO-CH2-CH(NH2)-COOH
Aspartic acid	asp d	HOOC-CH2-CH(NH2)-COOH
Cysteine	cys c	HS-CH2-CH(NH2)-COOH
Glutamic acid	glu e	HOOC-(CH2)2-CH(NH2)-COOH
Glutamine	gln q	H2N-CO-(CH2)2-CH(NH2)-COOH
Glycine	gly g	NH2-CH2-COOH
Histidine	his h	NH-CH=N-CH=C-CH2-CH(NH2)-COOH
Isoleucine	ile i	CH3-CH2-CH(CH3)-CH(NH2)-COOH
Leucine	leu l	(CH3)2-CH-CH2-CH(NH2)-COOH
Lysine	lys k	H2N-(CH2)4-CH(NH2)-COOH
Methionine	met m	CH3-S-(CH2)2-CH(NH2)-COOH
Phenylalanine	phe f	Ph-CH2-CH(NH2)-COOH
Proline	pro p	NH-(CH2)3-CH-COOH
Serine	ser s	HO-CH2-CH(NH2)-COOH
Threonine	thr t	CH3-CH(OH)-CH(NH2)-COOH
Tryptophan	trp w	Ph-NH-CH=C-CH2-CH(NH2)-COOH
Tyrosine	tyr y	HO-p-Ph-CH2-CH(NH2)-COOH
Valine	val v	(CH3)2-CH-CH(NH2)-COOH

In 1960 Marshell W. Nirenberg began to collaborate with Johann Matthaei at the National Institutes of Health. They prepared an extract from bacterial cells that could make protein even when no intact living cells were present. Adding an artificial form of RNA, polyuridylic acid, to this extract caused it to make an unnatural protein composed entirely of the amino acid phenylalanine (F). This provided the first clue to the "code" through which RNA (and, ultimately, DNA) controls the production of specific types of protein in the living cell.

Nirenberg announced these results at the International Congress of Biochemistry in Moscow, August 1961. Not everyone present realized the momentousness of the occasion.

Over the next few years many similar experiments were done, by Nirenberg and others, using different sequences of synthetic RNA to stimulate protein production. However, only modest further progress could be made in deciphering the code by such methods. But in 1964 Nirenberg announced that he and Philip Leder had devised a new and more powerful decoding technique, and within a year the code was fully deciphered.

The codon TTT codes for Phe (F), TTA codes for Leu (L), TCT codes for Ser (S), TAT codes for Tyr (Y), TGT codes for Cys (C), TGG codes for Trp (W), CTT codes for Leu (L), CCT codes for Pro (P), CAT codes for His (H), CAA codes for Gln (Q), CGT codes for Arg (R), ATT codes for Ile (I), ATG codes for Met (M), ACT codes for Thr (T), AAT codes for Asn (N), AAA codes for Lys (K), AGT codes for Ser (S), GTT codes for Val (V), GCT codes for Ala (A), GAT codes for Asp (D), GAA codes for Glu (E), and GGT codes for Gly (G).[42] *In general, more than one triplet codes each amino acid, so this list is simplified by not including alternate codons for the same amino acid.*

The DNA within the nuclei of cells makes up the chromosomes. There are twenty-three pairs of chromosomes in all human cells, except ova and sperm cells which have only one member of each pair. A gene is a small segment of a chromosome. A gene may contain a sequence of codons that determines the sequence of an enzyme (enzymes are proteins that control chemical reactions) or a structural or functional protein (like hemoglobin, myoglobin, or insulin). A tiny sequence of a gene may read this way:

taaacttcatggcataaccttgccaaagtatactaagaataaccctgaca
caaagctctttttttcagccaacatgccatgaaagaaagaagacaagggggt

gatctccactctctaagtgaaccactaaacccaccaaagaagaaacgagg
gaaatagaaagaggacccttgcctgagataatggatctgtatgtatgagt
agtagaaccctgctcaaagtacaaggaagggaaaaaaaagttagtttatt
tggaattttggacattaagagtctttattgttcattttcttttaactcac

(Short sequence from chromosome 1: bases 215847819–215860406 [reverse complemented] located in 1q31.1.)

Almost every cell in your body has a copy of all your genes. It would be chaos if the gene that specifies for insulin synthesis, for example, were to be expressed in your urinary bladder, or the gene for nail growth were expressed in your eye. That's where on/off switches, or gene promoters, come into play. On/off switches sit upstream in the reading frame from the genes that they regulate and dictate whether, or to what extent, those genes are turned on or off. Some switches, such as TATA boxes, are easy to recognize. They usually have a sequence such as TATAAA which allows certain proteins in the cell to attach to that piece of DNA and turn on the gene. Other on/off switches are harder to find and less understood.

The entire sequence of all the human chromosomes has now been worked out (see advance 74). It contains about three billion base pairs. With this information, human structure, human nature, human health, and human culture can all be altered. How this sequence of bases evolved is rapidly being clarified. What to do with the power inherent in this knowledge is being discussed, even fought over. Simply put, the argument centers around this question: Who owns the human genome?

As mentioned, the US Patent and Trademark Office (PTO) is awarding patents on the cellular structure of living organisms. With these patents businesspeople have barred traditional cultures from using their beans, their colored cotton—products used for hundreds of years—unless they pay license fees. Now, businesspeople, wearing the hats of scientists, have patented portions of the human genome and are charging researchers and others for license fees, claiming that these sequences are their property.

Three principle arguments have been articulated against DNA sequence patents:

(1) Patents should be granted only to inventions, not on something that is a discovery of nature. DNA sequences are a discovery of nature and, therefore, are not patentable.

(2) DNA sequence patents potentially restrict research in particular areas and may inhibit progress in medical research in the treatment of a disease.

(3) DNA sequence patents are morally objectionable because the patent creates a property right in the building blocks that make up humankind.

The opposite side argues:

(1) The effort involved in locating, characterizing, and determining the role genes play elevate the discovery of their sequences to the status of an invention, not merely a discovery.

(2) Discoveries of this nature are expensive, in terms of time and money; thus, obtaining a patent may be the only way that companies or organizations can protect their investments.

(3) Patents may promote research and development because patents facilitate focusing of effort and inhibit duplication of research.

In our opinion, these last three arguments are specious. It may have been difficult to get to the moon, but that did not allow those who did to claim discovery or ownership. And those who discovered the genetic code, along with the vast majority of other discoverers in biological and medical science, were not inspired by or funded by corporate money. It's one thing to privatize social security or the armed forces, but it's another thing to privatize human life. The stakes are astounding, the race is on, the deck is stacked, and the public needs to be informed.

According to Linda Pannozzo, in 2000 a British woman was the first person in the world to try to do something seemingly unthinkable: "Donna Rawlinson MacLean tried to patent herself. Her application for the patent was titled 'Myself' and her reasons for trying to do it were as good as any."[43]

"It has taken 30 years of hard labour for me to discover and invent myself, and now I wish to protect my invention from unauthorized exploitation, genetic or otherwise," MacLean told the British newspaper the *Guardian*. That "unauthorized exploitation" MacLean was talking about is quickly becoming a force to be reckoned with, as genomics companies hustle to profit from the bits of information that come together to form a blueprint for life—our genetic code. And the potential profits are staggering. Sales of DNA-based products and technology are projected to exceed $45 billion by 2009.

In 1988 a transgenic mouse containing genes from chickens and humans was patented by DuPont. In 1998 Iceland's Parliament created the Icelandic Health Sector Database Act selling the genetic heritage of the heirs of the Vikings to the Reykjavik-based genomics company deCODE Genetics Inc., who turned around and signed a five-year access agreement with Hoffman–La Roche for $20 million.[44]

In our opinion, research in these areas and those of gene therapy should proceed under close scrutiny and regulation by wide and diverse segments of the population. This is about the future of life.

Suggested Reading

1. Francis Crick, Nobel lecture, December 11, 1962, "On the Genetic Code," *Nobel Lectures: Physiology or Medicine 1942–1962* (Amsterdam: Elsevier, 1967).

2. M. W. Nirenberg and J. H. Matthaei, *Proc. Natl. Acad. Sci.* U.S. 47 (1961): 1588.

3. M. W. Nirenberg, J. H. Matthaei, and O. W. Jones, *Proc. Natl. Acad. Sci.* U.S. 48 (1962): 104.

4. Marshall W(arren) Nirenberg in *Current Biography*, 1965.

5. Horace Freeland Judson, *The Eighth Day of Creation: The Makers of the Revolution in Biology* (New York: Simon & Schuster, 1979).

Tropical Medicine

Take up the White Man's burden—
Send forth the best ye breed—
Go, bind your sons to exile
To serve your captives' need.

—Rudyard Kipling (1865–1936)

Now the long-feared Asiatic colossus takes its turn as
world leader, and we—the white race—have become the
yellow man's burden. Let us hope that he will treat us
more kindly than we treated him.

—Gore Vidal (b. 1925)

Take me up rivers of fear into the heart of darkness, where the white man searches for treasure, slaps at mosquitoes, and curses underdevelopment. It may be surprising to learn that during the nineteenth and early twentieth centuries malaria was endemic in many areas of the United States. Malaria control today requires a global strategy of socioeconomic sustainable development, and its history, like that of tropical medicine, is among the more revealing stories of Western culture.

Here are the names: Manson, Bancroft, Burnet, Derrick, Wucherer, Low, Lewis, and Ross—men from Scotland, England, Germany, and Australia, whose lives and work with filariasis, malaria, yellow fever, amebiasis, loa loa, paragonimaiasis, and schistosomiasis are the history of what Westerners call tropical medicine. It is a history of science, personal sacrifice, and achievement, but also of conquest and colonialism. And it is still evolving, so that we may shape the tragic story of tropical disease to a beneficent conclusion.

Tropical diseases afflict hundreds of millions of people, causing terrible human tragedy and death when poor living and working conditions allow parasites to run rampant. In large measure, these conditions have been perpetuated by Western colonialist exploitation, and the stimulus to study and control them has been the protection of Westerners going into areas where they are endemic.[45]

Above all others there was Patrick Manson, the "father" of tropical medicine, whose story tells almost all. The son of a bank manager who was lord of Fingansk, an estate near Aberdeen, Scotland, he studied medicine and qualified in 1865 to become a doctor at the age of twenty-one. His mother was a cousin of David Livingstone, the celebrated medical missionary and African explorer, while her father, Patrick Blaikie, was a naval surgeon who served aboard the HMS *Undaunted* when it took Napoleon to exile in Elba in April 1814. In 1866 Manson was appointed medical officer for Formosa (now Taiwan) in the service of the Chinese Imperial Maritime Customs, and after five years he took a similar appointment at Amoy (Xiamen). During Manson's first nine years in China, he worked in missionary hospitals and saw many cases of elephantoid deformity of the scrotum and lower limbs: he called it lymphscrotum. In 1875, while on leave in Britain and studying in the library of the British Museum (sharing its reading room with Karl Marx, who was there writing *Das Kapital*), he came across the writings of Timothy Lewis, also a graduate of Aberdeen and an officer in the Army Medical Services in India. Lewis had described filarial worms in an East Indian that had caused the same disease Manson had seen in China. Manson returned to China with renewed interest in tropical disease and made several important contributions to the understanding of the disease caused by the worms (now called *Wuchereria bancrofti*), especially the suggestion that it is transmitted through the bite of a mosquito.

But more than anything he became convinced that the greatest threat to the health and survival of immigrants from northern temperate countries to the tropics was the high prevalence of endemic diseases and not the climate.

Manson would go on to make his greatest contribution in stimulating and advising Ronald Ross to validate his theory that mosquitoes transmit malaria. Nevermind that Susruta, a Brahmin priest, had described a malaria-like illness which he considered was spread by mosquitoes more than fifteen hundred years earlier. Of this, and other similar reports, they were certainly unaware. Ross got the second Nobel Prize for his work, and Manson, a modest man whose ideas and encouragement were behind Ross's triumph, went on to create the London and Liverpool schools of tropical medicine. In 1898 he published his *Manual of Tropical Diseases*, which is currently in its twentieth edition. Patrick Manson retired from the Colonial Office in 1912 and died in 1922 at the age of seventy-eight.

Today most tropical diseases are still rampant among indigenous peoples. For example, a global map of the deadliest form of malaria put out by the Communicable Disease Centers (CDC) shows there were 515 million cases of the disease in 2002. The figure is up to 50 percent higher than World Health Organization estimates and 200 percent higher for areas outside Africa. More than two-thirds of cases occurred in Africa, where the malaria organism *Plasmodium falciparum* mostly affects children under five. But far more people are infected in Southeast Asia than anyone previously suspected. In excess of two billion people around the world—about a third of the global population—may be at risk of malaria. These data were compiled by scientists from the University of Oxford who work at the Kenya Medical Research Institute in Nairobi. They based their findings on recent and historical epidemiological, geographical, and demographic data, and they recently published their work in the journal *Nature*. Effective coordination and cooperation between research and control coalitions under a global umbrella group is urgently needed.[46] While a vaccine is also needed, it is unavailable because of a lack of priority in the developed world. Nonetheless, the techniques and drugs are here to make the long-deferred dream of malaria control a reality. But that dream has remained a raisin in the sun for a long time.

30.

Microscopes

The first major commitment a prospective doctor makes to a life in medicine is to acquire a microscope. This has been so for over one hundred years. Usually you take it with you on the first day of medical school. (I remember it like yesterday, because it was my first test. It had rained hard, thoroughly soaking me, and I had to appear before the Microscope Committee of the faculty. I was carrying an old monocular microscope in a beaten up wooden box. I had warned my father that the committee's letter said they were strict, that they recommended binocular microscopes with various attachments. He found that highly amusing and proceeded to tell me all about Leeuwenhoek and that I should concentrate on what I'm thinking when using the scope, not on how old and worn it is. The committee was thrilled to see such an old Bausch and Lomb in good working condition.)

The earliest known mention of the term *magnifying glasses* is found in the writings of Seneca and Pliny the Elder, two first-century CE Roman philosophers. But it was nearly twelve centuries later before the technology of curved lenses, used in the first spectacles, was exploited consistently.

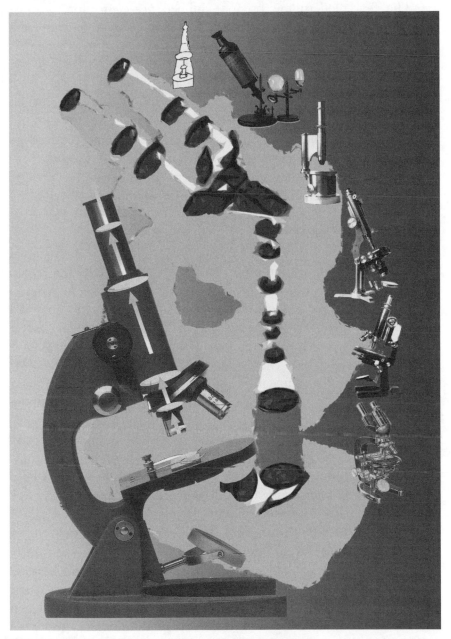

Fig. 26. The evolution of light microscopes with a representation of lenses and condensers within a microscope.

The first modern microscope, dubbed *flea-glasses* for its most common usage, was nothing more than a single curved lens inserted into a tube. This simple device captured the imagination of scientists across the globe, despite magnifying the object at the other end to just ten times its actual size. Then, in the late sixteenth century, two Dutch spectacle makers, Zaccharias Janssen and his son Hans, began experimenting with multiple lenses in a tube, inventing what would later develop into the increasingly powerful compound microscope and eventually the modern telescope. The Janssens' revolutionary advance was almost immediately greatly enhanced by Galileo in 1609, when the man, later considered the father of modern physics and astronomy, added a focusing device.

But it wasn't until Anton van Leeuwenhoek (1632–1723), a self-taught Dutch scientist, grinded out lenses of significantly increased curvature and magnification power (270X) that the microscope began to revolutionize science and medicine. He was the first to see and describe bacteria, yeast plants, the teeming life in a drop of water, and the circulation of blood corpuscles in capillaries. Today, using Leeuwenhoek's principles combined with modern instruments pioneered by American lensmaker Charles A. Spencer, compound microscopes can achieve magnifications up to 1,250X with normal light and 5,000X with blue light. And we are currently enjoying a revolution in light microscopy: differential interference contrast (DIC) microscopy, Hoffman modulation contrast, oblique or anaxial illumination, polarized light microscopy, Rheinberg illumination, stereomicroscopy, deconvolution, confocal microscopy, microinterferometry, near-field scanning optical microscopy (NSOM), and fluorescence microscopy techniques.[47]

Yet even with 5,000X magnification, there are an infinite number of important "microscopic" particles and life-forms that are utterly undetectable. A compound microscope, by definition, is reliant on light for illumination and cannot be used to distinguish objects that are smaller than half the wavelength of light. That means that two lines which are closer together than 0.275 micrometers cannot be seen separately, and any object with a diameter smaller than 0.275 micrometers will be all but invisible. To see these tiny objects, scientists must bypass light altogether and use an illumination source with a shorter wavelength.

In 1931 two Germans, Max Knott and Ernst Ruska, invented just such a tool: the electron microscope. Using a vacuum to speed up electrons until their wavelength is just 1/100,000 that of white light, an electron microscope can focus beams of those electrons on a cell until they are absorbed or scattered, in order to form an image on a electron-sensitive photographic plate. This process has the capacity to magnify objects up to 1,000,000X. This allows us to see viruses and the fine structure of cells. What it cannot do is help scientists observe the movements of living specimens because life cannot survive in a vacuum. For this type of work, modern biologists must revert back to the basic principles of Antonie van Leeuwenhoek.

31.

The Fool of Pest

Medical advancement is dependent upon people's inclination and ability to think new thoughts, test novel concepts, and accept ideas that may be contrary to established teaching. Ignaz Semmelweis (1818–1865) understood that. Not everyone does.

History is littered with the corpses of innocent victims, slaughtered by a doctor's stubborn mind. But very few of the intellectual battles sealing their fate played out as transparently as Semmelweis's nineteenth-century fight to exorcise the demons from his maternity ward.

About three decades before Pasteur championed the germ theory of disease (see advance 15), Semmelweis was a young assistant physician at the Vienna General Hospital, working in the largest maternity clinic in the world. He was assigned to Ward One, where the mortality rate from childbed fever, a septic condition occurring in mothers shortly after delivery (also called puerperal fever, or endometritis), was an astounding 30 percent. Nearby, in Ward Two the motality rate was only 3 percent. Semmelweis observed that the difference between the wards was that in Ward One the new mothers were attended by medical students and physicians who often arrived straight from the autopsy room without washing their hands. Ward Two was attended by midwives and midwifery students, who were more concerned with cleanliness and washed their hands after each examination. So Semmelweis switched staffing arrangements, and the high mortality followed the medical students to Ward Two. In 1847 he ordered that personnel wash their hands with chlorinated water before all deliveries. The mortality rate fell promptly to 1.27 percent.

Now, none of this was new. In 1795 Alexander Gordon wrote *A Treatise on the Epidemic Puerperal Fever of Aberdeen*, in which he argued that the cause of puerperal fever was the introduction of "putrid" (infected) material into the uterus by the midwife or doctor. He recommended washing. In 1842 the Boston gynecologist and author Oliver Wendell Holmes (1809–1894) was teaching these same principles in America, including hand washing with chlorinated water. He was ridiculed by the reigning gynecological authorities, led by Dr. Joe V. Meigs of Philadelphia. He argued that the disease was not infectious and that it was "ridiculous" to suppose that it could be spread by the hands of physicians, who after all were gentlemen. Holmes got nowhere with his hand washing efforts, though he did become the dean of Harvard Medical School.

But Semmelweis, five years later, stuck with it. He performed clinical experiments to prove the efficacy of hand washing, and he wrote letters to inform physicians of the importance of hand washing, and what he achieved was the near-total disdain of his peers. He was forced to quit Vienna. He set up in Pest, Hungary, where he drove the childbed fever mortality rate to below 1 percent and achieved the same result with his colleagues. His data challenged their professional practice, their valuable image as unquestioned authorities. They called him "The Fool of Pest."[48] He called physicians who refused to follow his hand washing methods "murderers." In 1865 he died in a mental asylum. And now there is a hospital in Vienna named after him, and books and movies (*The Cry and the Convent*) sanctify him.

Despite Semmelweis's battles, it is surprising, nonetheless, that so many doctors, all of whom have been carefully selected and trained beginning in their most formative years, would so often lack the agility of mind commensurate with their vast body of knowledge. In the medical profession obduracy is a loaded gun. In fact, some scholars still claim that Semmelweis didn't contribute much. Today, hospital-acquired (nosocomial) infections are still frequent, and most medical workers, especially MDs, still do not wash their hands before examinations. The next time you find yourself in a hospital you can do a little experiment: record how often the sinks, which are provided in each room, are actually used. Then report your findings to the doctor in charge, and see what kind of reception you get.

32.

Contraction

I resolved that women should have knowledge of contraception. They have every right to know about their own
bodies. I would strike out—I would scream from the house-
tops. I would tell the world what was going on in the lives
of these poor women. I would be heard. No matter what it
should cost. I would be heard.

—Margaret Sanger (1879–1966), *My Fight for Birth Control*, ch. 3 (1931);
on the resolve she made in 1912 while serving as a
nurse to the poor on New York City's Lower East Side

The development of safe and effective contraceptive techniques represents the culmination of thousands of years of effort. The ability to control conception has allowed profound change in human behavior and belief. Contraception, like so much of medicine, involves issues that go to the core of our values, and thus is controversial. But the desire to control contraception goes back to the dawn of culture itself. Over the ages, men have endured some pain, but women have risked their lives in efforts to achieve contraception.

In Egypt the Kahyn papyrus (c. 1850 BCE) describes the use of pessaries of pulverized crocodile dung and herbs. Dried dung pessaries of many kinds were used in many places, although irritation and infection were both predictable and important for their efficacy. Less dangerous, and probably less painful, the snake skin condom

Fig. 27. Young Margaret Sanger with an egg covered by sperm, contraceptive pills, and a poster for a meeting in Carnegie Hall.

may, in this respect, represent a nadir of sensation and a pinnacle of male suffering.

The desperate race among millions of sperm for one winner to penetrate the ovum is so frought with difficult adventure that tinkering with the process in almost any way is likely to provide some measure of contraception. So the insertion of foreign material into the vagina, for example, a pessary, can interfere with the striving masculine masses in a number of ways. The presence of a foreign object can alter the motility of the female genital tract, confounding the precise timing of the fateful meeting of sperm and egg. The material itself, dung of various animals, fermented dough, acacia gum, can seal the cervical os with waxy material or by inflammation and infection close the entrance to the womb. Today's pessaries are called vaginal suppositories. They contain waxy material and spermicides such as nonoxynol-9. They must be inserted before intercourse.

Condoms, cervical caps, female condoms, and other physical barriers are now generally made of latex or polyurethane and used in combination with a spermicidal agent. Intrauterine devices (IUDs) are about three inches long, come in many shapes, and are frequently made of copper. Once placed in the uterus, they may prevent pregnancy for years by keeping the fertilized egg from implanting in the wall of the womb. Surgical procedures, like tubal ligation and vasectomy, prevent the passage of the male and/or female gametes.

Oral contraception goes back at least two thousand years. Queen Anne's lace, or wild carrot, is said to be highly effective and is among the first known oral contraceptives. Studies in mice show that it blocks the production of progesterone, though little scientific research has been conducted on herbal contraceptives and their side effects. Queen Anne's lace is still used in the rural United States as a morning-after contraceptive. Other plants used for contraception include silphium, pennyroyal, asafoetida, artemisia, myrrh, rue, willow, date palm, pomegranate skin, cabbage, juniper, pine, onions, and acacia gum. Some of these are toxic, but it should also be understood that specimens of these plants may vary greatly in potency.

Enter the Pill. Birth control pills are oral contraceptives for women. Since its introduction in 1960, the pill has helped women gain more control of their lives, and it has altered family life so profoundly that many consider it the most socially significant medical advance of the century. Consisting of progestin alone, or some combination of estrogen and progestin, current oral contraceptives are nearly 100 percent effective. Lower doses have maintained effectiveness and reduced dangerous side effects so that about eleven million American women currently use the pill, and it is the most popular method of nonsurgical contraception. Dr. Gregory Pincus developed the first estrogen-progestin–based birth control pill at the invitation of Margaret Sanger and the Planned Parenthood Federation of America. They contained significantly higher doses of the hormones and had higher rates of side effects including blood clots traveling to the lungs, heart attacks, and strokes. Before being introduced into the United States, these early pills were tested on six thousand women in Puerto Rico and Haiti.

Currently, there are numerous types of sex-hormone-based regimens to prevent pregnancy and regulate menses by altering the balance of ovarian steroids (primarily estrogen and progesterone), thereby preventing the pituitary gland's release of peptide hormones that stimulate ovulation. They also thicken the cervical mucus, and thus block the entrance of sperm into the womb, and interfere with the production of a nurturing lining of the womb, which is necessary for developing a fertilized egg.

The pill gave women veto power over their pregnancies, and men a way to abdicate responsibility. Hugh Hefner, publisher of *Playboy* magazine, who, in the wake of the pill, played the Sexual Revolution like a virtuoso, declared that it "changed sex from procreation to recreation." The pill can be taken in privacy, so the male partner can simply assume that the female is taking precautions. So, if the pill empowered many women, it also liberated many men from the burden of thinking about the consequences of their actions. Moreover, barrier methods and spermicide require a certain degree of cooperation, or at least awareness, regarding pregnancy precautions.

The idea of consequence-free sex certainly altered the dynamics of life for the sexually active, putting new pressures on concepts of sexuality, commitment, romance, and marriage. And while many of the attitudes and behaviors brought by the pill will continue to shape societies, the reemergence, with a vengeance, of sexually transmitted diseases, especially HIV/AIDS, has once again changed the mix of issues so that the young people of today can hardly adopt a casual attitude toward their sexual activities.[49]

33.

George Seldes Leads the Fight to Deal with Tobacco-Related Illness

Smoking . . . is downright dangerous. Most people who smoke will eventually contract a fatal disease and die. But they don't brag about it, do they? Most people who ski, play professional football or drive race cars, will not die—at least not in the act—and yet they are the ones with the glamorous images, the expensive equipment and the mythic proportions. Why this should be I cannot say, unless it is simply that the average American does not know a daredevil when he sees one.

—Fran Lebowitz (b. 1950), a dedicated cigarette smoker, who often writes satirically about smoking, laws restricting smoking, and "smokers' rights"

The American Medical Association contends that smoking-related illness is the leading cause of death in the United States. Four hundred thirty-five thousand people die every year as a direct result of smoking, a staggering 18.1 percent of all fatalities. That's twenty-five times more people than are killed (directly or indirectly) by all illicit drug use combined.[50]

So why is such a dangerous product legal?

The answer is a combination of money, political influence, tradition, and misinformation on the part of the tobacco industry dating back two hundred years. George Seldes, an internationally respected journalist born in 1890, spearheaded the effort to bring to light this dangerous combination of forces, which he believed started with a conscious manipulation of the media. The startling fact that the pioneering work of an American born two centuries ago continues unfulfilled to this day is a testament to the power of the tobacco industry.

Today, the tobacco industry's annual sales are estimated to be between $300 billion and $400 billion. And they continue to spread that money around with highly effective precision. Big tobacco spends $6.8 billion on advertising and promotions every year and over $65 million lobbying Congress on various issues. In 1998, when Senator John McCain sponsored a tobacco control bill, the tobacco industry spent $70 million to successfully fight against the bill on state ballots.[51]

Advertising and political lobbying are both legal and appropriate means for any industry to disseminate its views. Ignoring scientific findings is not.

Despite the well-documented medical dangers of smoking, the tobacco industry and its advocates have saturated society with misleading information. The Cato Institute—a libertarian think tank, which lists Philip Morris and R. J. Reynolds among its biggest financial backers[52]—is just one of the highly effective outlets. Take this dandy as a particularly sinister case in point: "The aim of this paper is to dissect the granddaddy of all tobacco lies—that smoking causes 400,000 deaths each year," declared an article ironically titled "Lies, Damned Lies" in the institute's magazine *Regulations*. "To be blunt, there is no credible evidence that 400,000 deaths per year—or any number remotely close to 400,000—are caused by tobacco. Nor has that estimate been adjusted for the positive effects of smoking—less obesity, colitis, depression, Alzheimer's disease, Parkinson's disease and, for some women, a lower incidence of breast cancer."[53]

In 1999 Philip Morris funded a study which concluded that early death was just one of the "positive effects" of smoking, since it would save the Czech government $146 million on healthcare, pensions, welfare, and housing for the elderly.[54] According to *ABC World News Tonight*, a California Superior Court jury found that both Philip Morris and R. J. Reynolds had acted with malice, knowing the health hazards of smoking and deliberately misleading the public about those dangers.[55]

This is exactly what Seldes warned against nearly a century ago.

Born in New Jersey at the turn of the twentieth century, Seldes was a freethinker of the highest order. By the 1920s he was a respected foreign correspondent for the *Chicago Tribune* among other papers, boasting exclusive interviews with Paul von Hindenburg, Vladimir Lenin, and Benito Mussolini. But his books—*Freedom of the Press* (1935) and *Lords of the Press* (1938)—detailing big money's stifling influence on American journalism did not play well in the McCarthy era and he was eventually forced, in 1940, to begin publishing an independent weekly titled *In Fact: An Antidote to Falsehoods in the Daily Press.*

The January 13, 1941, issue included his first of more than fifty reports on the fraudulent and dangerous practices of the tobacco industry. Pointing to a Johns Hopkins University study from three years earlier that documented the mortal dangers of cigarette smoking, Seldes attempted to connect the dots between tobacco—"an industry which pays the press $50,000,000 a year"—and the absence of coverage of the Hopkins study by the press, whose readers he reminded the world had a vested interest in such information since they were "the consumers of 200,000,000,000 cigarettes a year."[56]

"No publication in America, outside of scientific journals, told the whole story of how tobacco shortens human life," he reported. "In New York the *Herald Tribune, Sun, News, Mirror, Post,* and *Journal-American* suppressed this story, although the Associated Press, United Press and Hearst's International News sent it out, and although science reporters turned in stories."[57]

This came at a time, it should be remembered, when even scientific evidence did not discourage the American Medical Association from passing out free cigarettes during its conferences and advertising cigarettes in its journal (*JAMA*). Seldes was the first to publicize a clause in the tobacco companies/newspapers contract stipulating that "no news or/and no adverse comments on the tobacco habit must ever be published." And *In Fact* was one of the only outlets to report the documented inaccuracies in cigarette advertising such as the claims in the Lucky Strike ad that "20,679 physicians have confirmed the fact that Lucky Strike is less irritating to the throat than other cigarettes. Many prominent athletes smoke Luckies all day long with no harmful effects to wind or physical condition."[58] And this one from 1939, a year after the John Hopkins findings were released: "Philip Morris—a cigarette recognized by eminent medical authorities for its advantages to the nose and throat."[59]

Yet, despite Seldes best efforts, many of the very same tactics are being used by both the tobacco industry and their willing accomplices in the media and government to this day, to dire consequence. According to the American Lung Association, the tobacco industry targets, through advertising and promotion, 1.63 million new smokers a year to compensate for those who quit or die, concentrating on young people, women, and minorities.

In October 2002 the World Health Organization reported that 4.9 million people die each year of tobacco-related illness and warned that its projection of 10 million deaths annually by 2030 was too low.

"That the industry knew of the addictiveness of nicotine and perpetuated that addiction through manipulation of nicotine is clear from the documents we reviewed," wrote Dr. Richard Hurt of the Mayo Clinic in Rochester, Minnesota, and Dr. Channing Robertson of Stanford University in an article titled "Prying Open the Door to the Tobacco Industry's Secrets about Nicotine," published in a 1998 issue of *JAMA*. They noted that these documents are available to the public on the Internet at http://www.mnbluecrosstobacco.com or at the Minnesota Depository.

Their article was based on a review of internal industry literature made public during a 1994 lawsuit filed by the state of Minnesota against the Brown and Williamson tobacco company. Hurt and Robertson found that after smoking was first linked to cancer in the early 1950s, tobacco executives moved quickly to create "a strategy of creating doubt and controversy over the scientific evidence."[60] In one 1954 document, for example, an industry official wrote that future advertising campaigns must aim "to free millions of Americans from the guilty fear that is going to arise deep in their biological depths . . . every time they light a cigarette."[61]

The tobacco companies, with deep ties to the drug companies, used their leverage to curtail advertizing of products that aim to curtail nicotine addiction. It has been documented that tobacco companies were aware for many decades that their products were addictive and caused disease and death. On April 14, 1994, when the CEOs of the seven leading tobacco companies testified under oath before Rep. Henry A. Waxman's (D-CA) Subcommittee on Health and the Environment, they each swore they did not believe that nicotine is addictive. In fact, their companies had rigged the nicotine content of their products to addict people quicker and harder. But the men were smiling that day, they showed no sign of anxiety, they looked like they were out for a walk in the park. And they were charged with no crime. No, not a thing, and the joke is on us. Their companies have civil judgments against them. Just a little money when measured against their crimes, their profits, and the human loss. They have already made it back. No one went out of business. They are back at the old stand, and we are made to understand that money talks loudly and directly to our government and that we have legalized corruption and called it lobbying.

34.

Penicillin

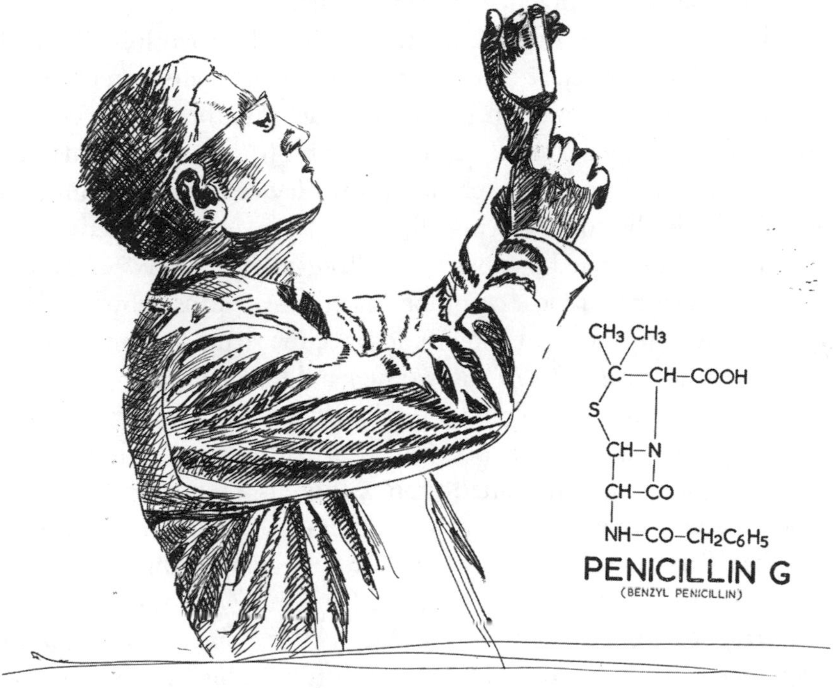

CH3 CH3
C—CH–COOH
S
CH–N
CH–CO
NH–CO–CH2C6H5
PENICILLIN G
(BENZYL PENICILLIN)

Fig. 28. Sir Alexander Fleming (1885–1955) discovered penicillin, the first antibiotic, in 1928.

A common theme that emerges among the greatest medical advances is the rediscovery of ancient or traditional knowledge and remedies. A few examples include smallpox vaccination (advance 16), the circulation of blood (advance 10), aspirin (advance 17), the idea of healing (advance 1), the relationship between healer and sick person (advance 2), the constant forgetting of traditional nutritional wisdom, and here, in the discovery of the antibacterial properties of molds. The triumph of penicillin, which flowered in the first half of the twentieth century, is rooted in folk medicine and its efficacy has now been nearly lost. This is a cautionary tale.

Traditional healing practice in many cultures included application of fungi to wounds or cuts. But in 1875 the English physician John Tyndall placed several tubes of clear culture broth around a room in an effort to determine the distribution of bacteria in a space. He then observed that those which developed a thick surface growth of Penicillium remained clear and that the bacteria which had grown lay dead or dormant as a sediment on the bottom. Having discovered the antibiotic activity of Penicillium, he published a paper in the *Transactions of the Royal Society*. But being unaware that most common infections are caused by bacteria, he made little of it, and so he did not advance the struggle against infection beyond those practices of wisewomen and folk healers.

At the same time, several observers, including Pasteur, became aware that when more than one bacterial pathogen is living in the same space (including on human tissue), there is a tendency for them to compete and for one to drive the others out, in accordance with Darwinian survivalist notions, which were then capturing the imaginations of the informed world. (Chares Darwin's *The Origin of Species by Means of Natural Selection: Or, the Preservation of Favored Races in the Struggle for Life* was published by John Murray Publishers of London in 1859.) In the microscopic struggle for existence, one life-form produced an "antibiotic" (against life) to vanquish others.

Alexander Fleming (1888–1955) was a Scottish bacteriologist working in London for decades on the ubiquitous problem of staphylococcus infection. He had already discovered the enzyme lysozyme, when in 1928 he accidentally "discovered" penicillin. In his lab at St. Mary's Hospital he left petri dishes growing cultures of staphylococcus and went on a short vacation. When he returned, he found that a mold that had grown on a petri dish had destroyed colonies of staph. If only Tindall had futher pursued his research! A half century wasted, millions of lives lost in suffering.

Fleming, of course, knew that staphylococci caused common diseases of the skin, wound infections, fatal pneumonia, and septicemia. He discovered that the antibiotic substance, which he called penicillin, was dissolved in the broth upon which the Penicillium mold grew. Moreover, he had found that unlike the disinfectant chemicals which he had studied earlier, penicillin did not damage the infected tissue or the inflammatory white blood cells and antibody molecules recruited to fight infections. In addition, he found that penicillin had antibiotic activity against most "Gram +" bacteria (those bacteria staining blue with Gram's stain, including streptococci, pneumococci, and diphtheria bacillus). But he never tested penicillin against syphilis, although he was considered England's leading expert in the treatment of this scourge, which the English called the French disease, and the French called the English disease. Syphilis was then being treated with a series of weekly intravenous injections of salvarsan (see advance 22) lasting for eighteen months. If he had tried oral or injected penicillin, he would have discovered its miraculous curative powers.

In fact, the great saga of penicillin is largely one of opportunities missed and squandered. Fleming wrote a few papers on his findings and then abandoned his work on penicillin, returning forever to his first love, lysozyme, the enzyme he had isolated from his own nasal mucus. It seems that the technical problems involved in culturing the mold and purifying penicillin were too much for him. Nasal mucus was readily at hand. The scientific community paid little attention to penicillin for the next ten years until Howard Florey (1898–1968) and Ernst Chain (1906–1979) (a refugee from Nazi Germany), working at Oxford, found Fleming's 1929 paper, and in 1940 they did a little experiment in which eight mice were injected with a lethal dose of staphylococci while only four of the eight also received penicillin. The next morning the four who had gotten penicillin were alive and the others were dead. The scientists were off and running.

The group led by Florey and Chain set up a production department at Oxford. They improved the methods of producing penicillin and studied the efficacy and safety of oral and intravenous treatment with it. They declined to apply for a patent, largely because of Florey's ethical principles. But Florey was well aware that large-scale commercial methods would be required to produce and distribute the quantities of penicillin that would be needed. Though it was 1941 and World War II was raging in Europe, Florey could not interest the British pharmaceutical industry in penicillin. A firm called Imperial Chemical Industries did produce enough so that he could treat a few dozen patients and establish dosage guidelines. Still, during the whole of World War II, British military and civilian penicillin needs were met by "mom and pop" bathroom and basement operations and a few small-scale manufacturers like Kendall, Bishop.

The impact of penicillin on the war effort and on the health of millions in the postwar era is clear. Florey came to the United States in 1941 to initiate American production, since the British, after all, were already under attack. The US pharmaceutical industry, starting with Florey's molds and methods, withheld important new information about production, causing hardship in Britain.[62] After cutting its teeth on the penicillin business, the US pharmaceutical industry became a major influence in every doctor's office, every living room, family farm, agribusiness, and government agency. According to the *New York Times*, even the FDA has strong ties to and performs only superficial monitoring of the pharmaceutical industry.

Unfortunately, largely because of the overuse of antibiotics, we are now losing their benefits (see advance 65).[63]

The Development of
Surgical Anesthesia

Give me to drink mandragora. . . .
That I might sleep out this great gap of time
My Antony is away.

—William Shakespeare, *Antony and Cleopatra*

Every body wants to have a hand in a great discovery. All I
will do is to give you a hint or two as to names—or the
name—to be applied to the state produced and the agent.
The state should, I think, be called "Anaesthesia" (from
the Greek word *anaesthesia*, "lack of sensation"). This sig-
nifies insensibility . . . . The adjective will be "Anaesthetic." Thus we might say the state of Anaesthesia, or
the anaesthetic state.

—Oliver Wendell Holmes, in a letter dated November 21, 1846,
to William Thomas Green Morton

s everyone knows, pain is normal and natural. Without
it we are at great risk, since without sensitivity to pain,
we will not react with defensive reflexes or other means
of self-preservation. But everyone wants to control
pain; this has always been the case. The modern
Western medical tradition assigns the beginning of surgical anes-

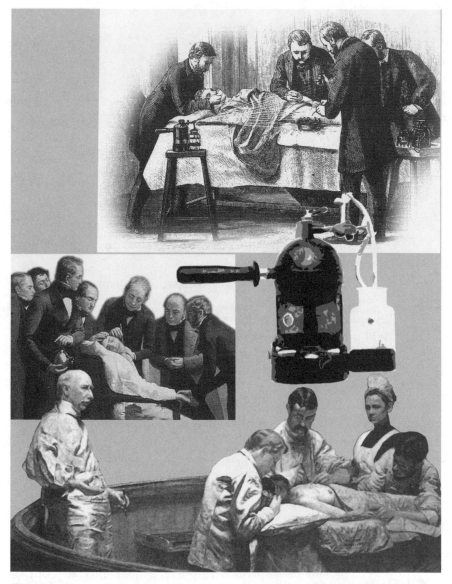

Fig. 29. A montage of images representing early operations conducted under anesthesia. In the center (*right*) is an early apparatus, the antiseptic carbolic steam spray.

thesia, however, to around 1846 (depending upon whose claim is recognized), and it is true that excellent pain relief, with loss of consciousness and memory, along with the muscular relaxation required for prolonged surgery, dates to that time. It entered with appropriate drama for such a great advance, for it was to make nonemergent (elective) surgery routine, give peace of mind and body to both patient and surgeon, and greatly expand the surgical repertoire.

But the goal of rendering people insensitive to pain, and its application to surgical practice, is probably as old as surgery. Some of the preparations used as surgical anesthetics were alcohol, cocaine, mandragora (extracts of the mandrake root, *Mandragora officinarum*), Indian hemp (hashish), and opium. These are well documented in ancient texts from around the world, along with instructions for their preparation and use. And the native peoples of South America gave us curare, in addition to cocaine, which is still widely used as a muscle relaxant, an important adjunct to surgical anesthesia. Still, the surgical use of these drugs was lost and found again during the course of Western history, and surgery retained much of its painful character even after 1846, as anyone with passing familiarity with our Civil War knows.

Before the 1840s, patients having surgery knew that they would have to experience pain. The toll that the pain, and its anticipation, extracted from all concerned with the procedure was often high and affected the performance of the operation. Often, the patient was premedicated with alcohol. Opiates such as laudanum were also used for this purpose, but these drugs left much to be desired in terms of pain relief and side effects. Then, the English chemist Sir Humphry Davy (1778–1829) found that nitrous oxide, a gaseous mixture of nitrogen and oxygen, could relieve his headaches and dental pain. Davy inhaled the gas and became euphoric; this was followed by uncontrollable outbursts of laughter, then sobbing, and finally loss of consciousness. He dubbed the chemical "laughing gas." Davy reported these experiences, but no one in the medical profession seems to have taken much notice.

At about that time laughing gas, and later ether, began to be used for recreational highs at posh parties. Queen Victoria had just married her cousin Albert. Charles Dickens was writing and agitating about social justice. Darwin was about to introduce evolution. In 1840s America there were traveling medicine shows where people got entertainment, a little "information," and a friendly rip-off. The salesmen in those days had something to grow hair on a billiard ball, something for ER (erectile dysfunction, then called lack of vigor), always with a hook line like "for erections lasting more than four hours . . .," and something to attract men, something to attract women, and "better" was always "better."

Horace Wells (1815–1848), a Connecticut dentist, wandered into such a show and witnessed a demonstration of nitrous oxide. Then, in 1844, he had a few of his own teeth extracted by a Dr. John Riggs under nitrous oxide administered by "Professor" Gardner Colton (1814–1898) at a fair in Hartford. Apparently he found that it helped because he organized a demonstration in Boston in 1845. But his patient screamed and struggled throughout the procedure, and poor Wells was ridiculed. Fortunately, one of the observers was another dentist, Wells's former partner, William Morton (1819–1868), who had begun experimenting with ether. Morton had no medical or dental degree, but had earned his stripes by apprenticeship, which was common. Morton constructed the first anesthetic machine, a simple device made of a glass globe housing an ether-soaked sponge. The patient just inhaled the vapor through one of two outlets. The use of ether was successful in Morton's dental practice, and the local newspaper began to publicize Morton's technique. Junior surgeon Dr. Henry Jacob Bigelow noticed these articles and arranged for a demonstration with his colleague, Dr. John Collins Warren, senior surgeon at the Massachusetts General Hospital. Young Morton, at age twenty-seven, was eager to gain notoriety. On October 16, 1846, in the surgical amphitheater of the Massachusetts General Hospital in Boston, a twenty-year-old man was successfully anesthetized by Morton so that a tumor could be painlessly removed from what one source said was his neck and another indicated was from his jaw.

But in Georgia they credit the first surgical anesthetic use of ether to Dr. Crawford Williamson Long, of Jefferson, Georgia. He claimed to have removed one of the two tumors from the neck of Mr. James Venable under ether anesthesia on March 30, 1842. He said he had used ether for minor surgery as early as 1841, and had originally learned about ether during "ether frolics" while in medical school at the University of Pennsylvania. Unfortunately for Dr. Long, he did not publish the results until 1848 in the *Southern Medical and Surgical Journal.*

By 1847 ether and chloroform were being used as general anesthetics for painful procedures in America and Europe. Childbirth was an exception because physicians worried about the effects of chemical analgesics on the fetus, and there were also concerns that the absence of pain would impair the bond between mother and child. The clergy fought anesthesia in labor because the Bible reports that God said to Eve, "in sorrow thou shalt bring forth children." So when Queen Victoria used chloroform for her eighth childbirth in 1859, there was shock and awe. She did it again for her ninth. The practice became widespread, until newer methods supervened, and now we have the return of "natural childbirth" (see advance 60).

There are now a host of gaseous and intravenous anesthetics, each with its own spectrum of potency, side effects, and toxicities. Add to that the array of sedatives, muscle relaxants, and other drugs used to make the operating room experience as free of trauma as possible. Now consider the advanced respirators, body temperature controls, monitoring equipment, and automated fluid infusion devices, and it becomes clear why anesthesiology is an advanced subspecialty and the operating room team must be well trained. Here is a general description of the process:

In a typical clinical procedure, known as balanced anesthesia, the patient is pre-medicated with a sedative intended to relieve pre-operative anxiety and facilitate the induction of anesthesia. Often this is a benzodiazepine such as diazepam or midazolam; otherwise, a barbiturate such as thiopental or nonbenzodiazepine such as propofol may perform this function. Sedation is followed by the induction of general anesthesia by intravenous injection of a sedative, narcotic (e.g., morphine, fentanyl, alfentanyl), or ketamine. In addition, a non-depolarizing curare-like derivative (e.g., vecuronium, d-turbocurarine) or a depolarizing drug (e.g., succinylcholine) is administered to induce muscle paralysis. After intubation and connection to a ventilator for artificial respiration, general anesthesia may be maintained by a mixture of oxygen and nitrous oxide, often in combination with a volatile agent (e.g., halothane, enflurane, or isoflurane) or intravenous drug. At the conclusion of the surgery, muscle relaxation is reversed (e.g., by neostigmine or other anti-cholinesterase), and normal (unassisted) breathing is restored. In addition, the patient may be given an analgesic agent (e.g., morphine) to manage any acute pain experienced postoperatively.[64]

This entry is focused on general anesthesia, in which the subject is unconscious. But there are other methods, such as "conscious sedation" and local or regional anesthesia, in which patients remain relatively or fully conscious. These are appropriate for specific procedures of various sorts, including major surgery. But general anesthesia is the great advance we celebrate, and we in medicine still don't know how it works. Neuroscientific research has suggested some hypotheses. It seems obvious that general anesthetics work on the central nervous system temporarily to inhibit synaptic transmission (the chemical means by which neural impulses are transmitted between adjacent neurons). This causes loss of consciousness and sensory awareness. But precisely where and how do the agents work? Perhaps the anesthetics impair the operation of the sodium pump (the enzyme that maintains the distribution of Na^+) across the cell membrane, preventing depolarization of postsynaptic neurons. Perhaps anesthetics simultaneously inhibit the actions of excitatory neurotransmitters such as glutamate and acetylcholine and enhance the actions of inhibitory neurotransmitters such as GABA and glycine. The specific physiological mechanism by which anesthesia is achieved probably differs for each class of anesthetic agents.

36.

Blood Transfusion

**Blood doubly unites us, for we share the same blood and
we have spilled blood.**

—Jean-Paul Sartre (1905–1980), Orestes to Electra in *The Flies*, act 2

William Harvey's demonstration that blood circulates in an essentially closed system (see advance 10) provided the context within which blood transfusion could be conceived and attempted. Within several decades an Oxford physician named Richard Lower was experimenting with dog-to-dog and animal-to-human transfusions. At various times and places during the sixteenth and seventeenth centuries, transfusions of blood, milk, and saline were attempted, but were met with unpredictable and frequently disastrous results.

The twentieth century rode in on a wave of microbiological investigation that revealed innumerable facts about specific infections and, perhaps more important, the complexity of human cellular (white blood cells) and antibody responses. Called the Age of Immunology, among its revolutionary advances was the development of blood transfusion. The great breakthrough came in 1901 when an Austrian physician, Karl Landsteiner, who was studying failed transfusions, found that mixing small amounts of blood from different subjects often, but not always, resulted in agglutination, or "clumping" of the red blood cells. It is this agglutination that causes sickness and death when incompatible blood is transfused. Landsteiner thus concluded that it must depend on the presence or absence of two protein antigens, which he called A and B, on the surface membranes of the red cells. He then mixed small samples of blood from many of his colleagues and grouped them as group A or group B based on whether they were compatible or clumped. When Landsteiner found blood from people which did not clump when mixed with either group A or group B, he concluded that the red cell membranes carried neither antigen and placed them in group 0 (zero, which became the letter O). Mixing and observing, he soon discovered yet a fourth group, AB, in which the cells carry both antigens.

The floodgates of blood transfusion did not, however, burst open. It took another forty years and the discovery of more red cell antigens (Rh, M, N, P), the development of anticoagulants, the use of preservatives, the founding of blood banks, and the onset of World War II for the lifesaving potential of blood transfusion to be realized. With the widespread utilization of transfusion came the development of posttransfusion disease, particularly viral hepatitis, and later AIDS. This has since been nearly eliminated by screening donor blood using radioimmunoassay (RIA, see advance 41) and other methods.

A central concern of this book is how these medical advances relate to all of us, whom they belong to, and how their use reflects the most fundamental and operative human values. Now, we realize that the word *values* has recently degenerated into a political tool, but consider the following facts. Since before recorded history, blood has been held precious, helping to define our species, our "race," and our very individuality. In 1935 a German doctor, Hans Serelman, had a patient who desperately needed a blood transfusion. This was before there were blood banks (except in Leningrad where the first one was established in 1932), and the procedure was done by direct transfer from donor artery to recipient vein. Having no other compatible donor, Serelman donated his own blood. He received no praise for this lifesaving act, for the doctor was a Jew, and he was sent to a concentration camp for defiling the blood of the German race. (He would have been sent to a concentration camp anyway.) In the years that followed, Germany sought to eliminate the Jewish "influence" on medicine by, among other things, preventing over eight thousand Jewish doctors from practicing. German immunologic studies were devoted to the trivial and fruitless task of distinguishing Aryan from Jewish blood. The Nuremberg Blood Protection Laws severely limited blood available for transfusion because of the possibility of being charged with "an attack on German blood" if the donor was found not sufficiently Aryan.

Meanwhile, back in the United States, the army was segregated and the Red Cross refused to collect blood from black people. After Pearl Harbor the need for blood was so great that they began collecting it, but they labeled and processed blood from black people separately. In the late 1950s, Arkansas passed a law requiring the segregation of blood. Louisiana made it a crime for physicians to give a white person "black blood" without obtaining prior permission. How many blacks and whites suffered and died because of this pervasive thinking? Some of the same legislators who supported those laws, policies, and attitudes have led our two political parties and have made decisions on all manner of health issues, from who gets healthcare, to the distribution of organs for transplantation, to abortion rights, to stem cell research, to who gets access to medical records, to who owns the human genome, and to the quality of our water and air.

Modern Surgical Techniques

Let us go then, you and I,
When the evening is spread out against the sky
Like a patient etherized upon a table.

—T. S. (Thomas Stearns) Eliot (1888–1965),
The Love Song of J. Alfred Prufrock

The development of modern surgical techniques required an atmosphere in which surgeons could operate for a prolonged period on a relaxed patient who would survive to recover. This situation had its beginnings in the latter half of the nineteenth century with the introduction of surgical anesthesia (see advance 35) and carbolic acid wound dressing (see advance 46). Under these conditions surgeons became free to develop their clinical virtuosity and to invent ingenious operations applying all the knowledge of anatomy and physiology that other physicians of that era could, for the most part, just talk about. The age of surgery was at hand.

Anesthesia had been in use for about fifteen years when Joseph Lister introduced his antiseptic techniques in 1865. Modern surgical techniques began to flourish in the last two decades of the nineteenth century and have improved since with new materials and methods. The old, sloppy operating theaters were replaced with glistening antiseptic operating rooms with metal surfaces. Rubber gloves, clean

162

starched gowns, and then masks and shoe covers were introduced. Electric lights, image-enhancing lenses, and suction devices have vastly improved the view of the operative field. A bewildering array of clamps, forceps, and other tools are at the surgeon's fingertips, but his most important asset is his knowledge.

With all the brilliant operative procedures that can now be done, some of the greatest advances in modern surgery don't take place in the operating room. They are the preparations and maintenance involved in pre- and postoperative care. The understanding of problems that can interfere with the patient's ability to withstand and recover from a surgery including the management of fluid and electrolyte (sodium, potassium, bicarbonate, etc.) imbalance, infection, and a host of other factors are crucial to the success of modern surgery. Also of vital importance is the judgment to know when, and when not, to operate.

38.

The Control of Polio

In 1955, some sixty years after the discovery of x-rays shook the world as no medical advance before, another advance captured the public's attention. This time the elements of television, a panicked population fearing for their children, and the vivid memory of a beloved savior (Franklin Delano Roosevelt) struck down by a dread disease added to the drama.

This great advance brought the greatest gift: freedom from fear. Not simply from "fear itself," but from fear of annual summer epidemics of paralytic poliomyelitis (then better known as "infantile paralysis"). In the five years before 1955, when mass inoculations with the new polio vaccine began, instances of paralytic poliomyelitis averaged about twenty-five thousand cases annually in the United States alone. Paralytic polio struck indivuals as far back as ancient Egypt. But in America, in the 1920s, 1930s, and 1940s, outbreaks of the disease increased in frightening epidemics. Many children and young adults died, were crippled, or became paralyzed, and they lived out their lives inside giant tubelike breathing machines called iron lungs. Experts expected the decade of the 1950s to be even worse, and in the epidemic of 1952, the worst on record, nearly fifty-eight thousand cases of polio were reported in the United States, and more than three thousand died of the disease. In addition to these many thousands of tragedies, summers were transformed into a time of high anxiety for everyone. No one who lived through that period can ever forget the fear.

By early 1955 the public was primed by the success of small trials of a vaccine against polio. Then, on April 12, Dr. Thomas Francis Jr. of the University of Michigan called a press conference to announce the success of a large field trial involving about one million subjects including some 440,000 children injected with Dr. Jonas Salk's "killed polio virus vaccine." It was truly a shot heard around the world and the joy and relief were palpable. President Dwight Eisenhower praised Dr. Salk as a "benefactor of mankind." The chairman of the board of directors of the American Medical Association, Dr. Dwight H. Murray, called it "one of the greatest events in the history of medicine."

Two crucial factors in the ultimate success of the polio vaccine were the earlier development of methods for growing viruses in the laboratory using animal-cell-tissue cultures and the finding that three strains of polio virus would have to be included to make an effective vaccine. Dr. John Enders of Harvard later won a Nobel Prize for that work. The Salk vaccine viruses were grown on monkey kidney cells, then inactivated (killed) by formaldehyde. After the successful field trials, Dr. Salk became a public hero. Opinion polls ranked him roughly between Churchill and Gandhi as a revered historical figure. Salk did not demur.

In his later years Dr. Salk became a philosopher. He came to believe that it was the force of evolution that guided him in his work and that he could guide evolution. "I have come to recognize evolution not only as an active process that I am experiencing all the time but as something I can guide by the choices I make, by the experiments I design," he said. "I have always sensed this as the next evolutionary step. It's not something of which a great many are capable, but some are." It is clear that Salk was a very proud man who was deeply hurt by the fact that, while he was greatly honored, the greatest honors eluded him, and many of his peers did not hold his work in high scientific regard. As Al Rosenfeld, the longtime science editor of *Life* magazine and a close friend of Salk put it, "The irony was that he was idolized by the public, and only a minority of his peers ever publicly recognized what he had done."[65] He never received a Nobel Prize and he was not elected to the National Academy of Sciences.

Dr. Julius S. Youngner was Salk's number one aide during the development of the polio vaccine. In a 1995 interview he was quoted as saying,

> I saw [Salk] as a father figure at the beginning and, at the end, an evil father figure. . . . He got the money and did the administrative things. He carried out the whole operation. And there was a lot of politics. But it took the stars out of my eyes when he was feeding stuff to the press and not really acting as though anybody else was involved. Everyone likes to get the credit for what they've done. He hid us. It took me a long time to catch on to that. It was a big shock when you realize that somebody that you admired and trusted had not done right by you and other people.[66]

Confronted with these and other remarks about his alleged self-aggrandizing, Salk responded, "Perhaps a more conscious attempt might have been made and perhaps should have been made to list the names of each individual more prominently rather than, as was implied, that the satisfaction came from the work itself. So, I can see that, given that the coin of the realm in science is recognition, that the focusing of attention on me—let me see how to put it. There's a heavy price one has to pay for that."[67]

But as much as some of Salk's colleagues objected to what they considered grandstanding and failure to share credit, there were much more serious scientific criticisms of his work and behavior. Most of this involved his rivalry with Dr. Albert Sabin and Sabin's live-attenuated polio vaccine.

Albert Sabin (1907–1993) was a brilliant virologist who, before World War II, had made important contributions in understanding the transmission of the polio virus. It was Sabin who discovered that polio gained entrance through the mouth and gastrointestinal tract before taking residence in nerves and destroying them. And during the war he advanced the fight against viral diseases that were weakening troops. For these achievements, Sabin was elected to the National Academy of Sciences in 1951.[68]

After the war he returned to his work with polio, making vaccines using very weakly pathogenic, essentially harmless, but live, strains of the three polio viruses ("live-attenuated vaccine"). Sabin theorized that the live-virus vaccine would have several important advantages over the killed-virus vaccine: it could be given orally in a lump of sugar, rather than by injection; one oral dose would result in lifetime immunity, rather than requiring several booster injections; and, like the active polio virus, the attenuated virus would be shed in the stool, thus infecting and protecting a wide circle of people around each immunized individual.

Salk campaigned against the Sabin vaccine. But Sabin was right about each and every point in his effort to provide the live-attenuated vaccine to the public. Still, Sabin was forced to go to the Soviet Union to test his vaccine in 1958 to 1959. This does not negate the fact that the Salk vaccine reduced the rate of paralytic polio to a few cases per year in the years between 1955 and 1962. But the excellent results of Sabin's live-attenuated polio vaccine in the Soviet Union earned it a trial in 180,000 schoolchildren in Cincinnati in 1960. It performed splendidly and was licensed by the United States Public Health Service in 1962, and since then it alone has been used in the United States and almost every other country in the world.

Asked about Salk in 1991, Sabin, famously outspoken, said, "Salk never discovered anything."[69]

39.

Cardiac Catheterization and Related Techniques

The heart of another is a dark forest, always . . .
—Willa Cather (1873–1947)

The heart is a drama queen: beating, racing, stopping, aching, or bleeding. And although the work of William Harvey (see advance 10) did much to demystify the heart, even today its *aura paradoxicus* of precious fragility and unfailing resilience rightly persists and it continues to be held in near-sacred esteem by folklore, philosophy, and medical science. So it comes as no surprise that ideas and attempts to tinker with the ticker have met with resistance on ethical, moral, religious, and scientific grounds. And that's why it took three hundred years from the time of Harvey to the next giant breakthrough in cardiovascular research and treatment.

A dispassionate view of the heart emphasizes its mechanical function as a pump. It has in-and-out flows, four pumping chambers, and four valves. Again, its basic functional anatomy was clarified as far back as 1628 in the publications of Harvey. Still, the precise measurement of pressure, volume, and flow rate within the chambers of the functioning human heart were the vital details required to define normal and diseased activity. And it was generally believed that entering the

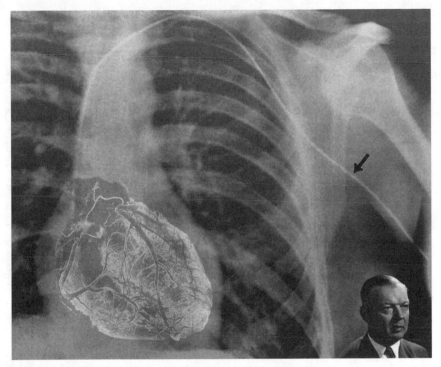

Fig. 30. Werner Forssmann (*lower right*) with his own x-ray showing the catheter (arrow) that he passed through his arm vein into his heart (superimposed on x-ray).

heart with an instrument would prove fatal. Nevertheless, in 1929 Werner Forssmann (1904–1979) did just that. He was a surgical resident at a small hospital in Germany, looking for a new method to administer drugs. So he inserted a ureteral catheter into his own anticubital vein (at the elbow), threaded it up sixty-five centimeters, and, with the catheter dangling from his arm, walked through the hospital to the basement x-ray room to document the catheter's location in his right atrium. For this courageous self-experimentation—a breakthrough that would lead to the new worlds of interventional cardiology and radiology—he was promptly fired from his job, denounced as a nut by the medical establishment, and almost completely ignored.

But in 1941, working at New York's Bellevue Hospital Chest Service, Dickinson Richards (1895–1973) and Andre Cournand (1895–1988) began a series of experiments using cardiac catheterization methods based on Forssmann's technique, and they revolutionized cardiopulmonary medicine, opening vistas of diagnosis and treatment, and winning the Nobel Prize for them and Forssman in 1956.

By placing catheters and pushing them through the body's vascular highways and byways, physicians can easily and safely reach, see, measure, and work almost anywhere in the human body. The results were initially diagnostic marvels: precise measurements of the hemodynamic relationships between the heart and lungs, calculations of valve areas to define stenosis (restricted opening) and incompetence (failure to close), and the injection of contrast materials to visualize structure and flow through organs and vessels (angiography). Then came therapeutic miracles like dilating stenotic vessels with balloons and stents in organs that were damaged by vascular disease and ablating the abnormal foci wihin the heart where life-threatening arhythmias had arisen. And now we restore vascular flow to prevent organ damage *before* it occurs in the heart, brain, kidneys, intestines, legs, almost anywhere. This was all made possible by Werner Forsmann's brave experiment and the brilliance of those who followed.

Today, hundreds of thousands of cardiac catheterizations are performed each year. They are done on infants, children, adults, and the elderly. But a critical young medical student might argue that cardiac catheterization has become a multibillion-dollar business, a bonanza with life-giving gifts that poses ethical issues.

In considering this, the American College of Cardiology concluded that

> [c]hanging practice patterns in medicine, including the expansion of both managed care and for-profit physician entrepreneurial ventures, have altered the relationships among physicians, patients, and payers, creating potential conflicts of interest for the physician in maintaining the patient's best interest. The availability of sophisticated yet costly diagnostic and therapeutic technologies has also created new challenges for physicians, who may now serve simultaneously as physician, inventor, and investigator of new therapies for vascular intervention. Government and regulatory authorities now seek greater assurances that physicians respect the best interest of the patient in their clinical practice.[70]

40.

Cholesterol and the Metabolic Basis of Cardiovascular Disease

Call the roller of big cigars,
The muscular one, and bid him whip
In kitchen cups concupiscent curds.

—Wallace Stevens (1879–1955), "The Emperor of Ice-Cream"

Cholesterol is the most highly decorated small molecule in biology. Thirteen Nobel Prizes have been awarded to scientists who devoted major parts of their careers to cholesterol. Ever since it was first isolated from gallstones in 1784, almost exactly 200 years ago, cholesterol has exerted a hypnotic fascination for scientists from the most diverse domains of science and medicine.

—Michael S. Brown and Joseph L. Goldstein, 1983 Nobel lecture

Cardiovascular disease (CVD) refers primarily to heart disease ("ischemic" coronary artery disease) and stroke. In these conditions, the arteries supplying the heart and brain are obstructed by plaques containing cholesterol. The result is ischemia, or decreased blood supply, which can lead to death of tissue (necrosis). In people with CVD, the same process takes place in most of the body's arteries, but because

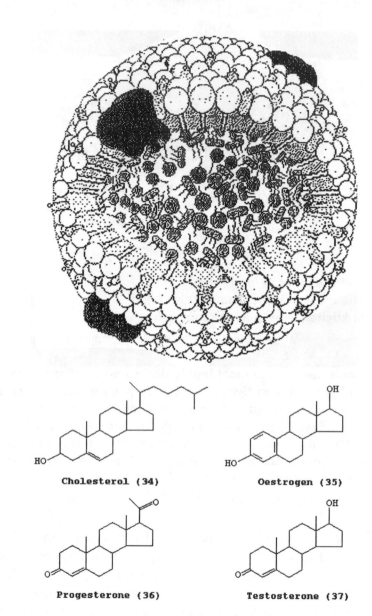

Fig. 31. A round particle of low-density lipoprotein (LDL), showing the arrangement of molecular constituents: apoprotein (3 dark surface patches), phospholipid molecules (outside), cholesterol ester (dark structures in the center), and triglycerides (center structures with three little tails).

of the vital functions of the heart and brain, the consequences are most often critical.

CVD is the nation's leading killer for both men and women among all racial and ethnic groups. More than one million Americans die of CVD each year. Heart disease is the leading cause of death for Americans age thirty-five and older. The death toll, while staggering, does not reveal the suffering of the fifty-seven million who struggle with the complications of CVD. Nevertheless, while much has been learned about the abnormal physiology and the risk factors associated with these conditions, no federal funding until 1998 had been directed to states to fight this disease.

Each year the estimated national costs of CVD increase. In 1999 they were estimated at $286.5 billion, including health expenditures and lost productivity. In fiscal year 1998 Congress made available funding of $8.1 million to initiate a national, state-based prevention program. This amount, together with the Centers for Disease Control (CDC) in-house funding of $2.8 million, totaled about $11 million and would have provided about $220,000 for each state to initiate a national, state-based prevention program. So they gave the funds to eight states: six (Alabama, Georgia, Kentucky, Mississippi, Missouri, and South Carolina) to build preliminary programs to target cardiovascular disease, and two states (New York and North Carolina) for more comprehensive programs. In fiscal year 1999, the CDC spent $15 million for the prevention and control of cardiovascular disease and its disabling conditions. This goes to the heart of our leading medical problem.

Cholesterol is a soft, waxy, crystalline substance distributed throughout the nervous system, skin, muscles, liver, intestines, and heart. Although high levels of cholesterol in the blood and other tissues are often associated with health problems, cholesterol is actually essential to human life. In molecular structure, it is a simple steroid, and it occurs in almost all animal and human fats, as well as in blood, bile, and cell membranes. It is produced primarily by the liver, but may be synthesized by other organs as well, and it may be introduced to the body in significant quantities through diet. Cholesterol is the precursor of vitamin D3 and a number of other types of steroids, such as bile acids and steroid hormones. The synthesis of the major steroid hormones from cholesterol occurs through the intermediary formation of a steroid known as pregnenolone, which contains the cholesterol nucleus but has a shorter fatty acid side chain. Pregnenolone is the precursor of progesterone, the female hormone that dominates during pregnancy, and of androgens or male sex hormones, such as testosterone. The estrogen female sex hormones, estrone and estradiol, and the adrenal corticosteroids, such as corticosterone, are also metabolic products of cholesterol.

Excessive cholesterol in the blood contributes to atherosclerosis and subsequent heart disease. The risk of developing ischemic coronary artery disease, or more broadly distributed atherosclerosis, increases as the level of blood cholesterol increases. Blood is basically modified sea water, and cholesterol is absolutely insoluble in water. So cholesterol circulates in the blood in association with soluble transport proteins. The lipid and protein amalgams form complex spherical structures of different densities, which include include HDL (high-density lipoprotein, commonly known as "good cholesterol") cholesterol; LDL (low-density lipoprotein) cholesterol; and VLDL (very low-density lipoprotein) cholesterol. Thus, the lipoproteins are spherical structures consisting of protein ("apoprotein," i.e., protein that combines with something else to form a conjugated protein), phospholipid (a lipid in which phosphoric acid as well as a fatty acid is esterified [attached] to glycerol and which is found in all cell membranes), cholesterol, cholesterol ester, and triglyceride. LDL, the most abundant cholesterol-carrying lipoprotein in human plasma, is arranged so that the particles (mass = 3×10^6 daltons; diameter = 22 nanometers) carry the lipophilic (fat-soluble, water-insoluble) molecules (primarily about fifteen hundred molecules of cholesteryl ester) in their oily core, while the water-soluble hydrophilic surface (composed of one molecule of a 387,000-dalton protein called apoprotein B-100, eight hundred molecules of phospholipid, and five hundred molecules of unesterified cholesterol) is exposed to the water-based blood. This allows the cholesterol and triglycerides to remain in solution. As the density of the lipoprotein increases, the relative amount of apoprotein rises, and the percentage of the lipids, which are lighter, decreases.

In 1972 two research scientists at the University of Texas, Michael S. Brown and Joseph L. Goldstein, set out to understand a genetic disease known as familial hypercholesterolemia (FH). Now also known by the descriptive names Type II hyperlipoproteinemia, hypercholesterolemic xanthomatosis, or low-density lipoprotein (LDL) receptor mutation, FH is characterized by markedly elevated LDL cholesterol levels beginning at birth and resulting in heart attacks at an early age. People affected by this dominantly inherited genetic condition have consistently high levels of low-density lipoprotein (LDL), which leads to premature atherosclerosis of the coronary arteries. In men with one FH gene (heterozygous), heart attacks occur in their forties to fifties, and 85 percent of men with this disorder have experienced a heart attack by age sixty. The incidence of heart attacks in women with FH is also increased, but delayed ten years later than in men. People with FH and other hyperlipidemias (high blood lipid disorders) frequently have obvious xanthomas (raised, waxy-appearing, frequently yellowish skin lesions, which are local deposits of the elevated lipid).

It is possible for a someone to inherit two genes for FH (homozygous). This intensifies the severity of the condition. Cholesterol values may exceed 6,000 mg/100 ml of blood serum. These individuals have thick, whitish (lipemic) serum and develop waxy plaques (xanthomas) beneath the skin over their elbows, knees, and buttocks. In addition, they develop deposits in tendons and around the cornea of the eye. Atherosclerosis begins before puberty, and heart attacks and death may occur before age thirty.

Brown and Goldstein saw this extreme condition as an opportunity to study the regulation of cholesterol synthesis. They hypothesized that cholesterol production might normally be regulated by a feedback process wherein cholesterol itself modulated its own production and that FH might result from a failure of end-product repression of cholesterol synthesis.

End-product inhibition of hormone release was then a well-known phenomenon. For example, the hormone gastrin, which stimulates acid production in the stomach, is prevented from further secretion by the end product, hydrochloric acid, as it lowers the pH (acidifies) the gastrin-producing region of the organ (the antrum). Similarly, secretin, which is produced in the duodenum (the first portion of the small intestine) and stimulates bicarbonate secretion by the pancreas, is inhibited by the bicarbonate as it enters the duodenum via the pancreatic duct. There are many such examples that were documented in the early years of radioimmunoassay (see advance 41). But in 1972, there was no data to support genetic defects in feedback regulation.

Brown and Goldstein used cell culture techniques to study the regulatory defect in FH. In this process they discovered a cell surface receptor for LDL. This receptor was found to mediate the feedback control of cholesterol synthesis—in other words, among its functions is the regulation of how cells make and utilize cholesterol. FH was shown to be caused by inherited defects in the gene encoding the LDL receptor. The receptor itself is a protein that snakes in and out of the cell through the cell membrane. This positioning, with extracellular (outside the cell), intracellular (inside), and transmembrane (through, or within, the cell membrane) domains, is precise and critical to the functioning of the LDL-receptor complex. After LDL binds to the cell-surface receptor, the complex is taken into the cell where it affects the regulation of lipid metabolism through the activity of enzymes such as 3-hydroxy-3methylguteryl coenzyme A reductase (HMG CoA reductase).

It is the understanding of the importance of diet, exercise, and the other risk factors mentioned above, along with the regulatory control on cellular cholesterol synthesis (e.g., inhibition of HMG CoA reductase by the "statin" drugs), that allows improvement of serum lipid levels, improvement of arterial pathology, and reduction of risk for heart disease and stroke. Indicative of cholesterol's central role is that this is effective despite the fact that we know there are still other factors involved in the process of atherosclerosis that we don't fully understand.

41.

Lab Rats Tango in the Bronx— The Development of Radioimmunoassay

Medical and scientific advancement is often dependent upon the ability to detect things and measure them with precision. The development of radioimmunoassay (RIA), a method that can measure infinitesimally small amounts of virtually any substance, revolutionized biomedical research and clinical medicine.

RIA, like a fine molecular microscope, can find and measure substances that previously had only been estimated using cumbersome and inaccurate biological assays. Solomon A. Berson and Rosalyn S. Yalow, of the Veterans Administration Laboratory in the Bronx, derived RIA through the study of the metabolism of insulin. Insulin is the hormone that lowers blood sugar. They had been tracing the fate of insulin by attaching radioactive iodine to molecules of beef insulin and injecting tiny amounts of the radioactive insulin into both normal subjects and diabetics. They were searching for an answer to one of the most important and perplexing problems in medical science: Why do most diabetics have high blood sugar if their insulin-producing cells are normal?

Some diabetics, who are frequently lean, develop the disease in

Fig. 32. Rosalyn S. Yalow and Soloman A. Berson, the developers of radioimmunoassay (RIA), which was first used to measure insulin.

childhood and rapidly lose the insulin-producing beta cells in the pancreas. Thus, they lose the ability to keep their blood sugar within the normal range. But there is another type of diabetes, one which was harder to crack. It was well known that the majority of diabetics, the ones who are generally overweight and develop the disease as adults, those who are now known as type 2 diabetics, have pancreata containing plenty of beta cells with lots of insulin. It was known that the insulin is normal in its ability to lower blood sugar and that it is released into the bloodstream as in normal people, after eating, as the blood sugar rises. Berson and Yalow wanted to see what happened to insulin once it entered the bloodstream in type 2 diabetics as compared to normal subjects. So they labeled trace amounts of insulin with radioactivity, injected it into human volunteers, people with and without diabetes, and they took blood samples every few minutes for several hours to determine how fast the insulin was metabolized and disappeared from the bloodstream. They found something completely unexpected, something that they hadn't even considered. In diabetics who had never been treated with insulin, the radioactive insulin circulated in the blood and was metabolized just like in normal people. But in people who had previously been treated with injections of beef or pork insulin, the radioactive insulin became attached to a large protein in the blood plasma. It was as if, in the treated subjects, the small active insulin molecule, like a little child, was taken into the arms of a sheltering adult and ushered through the bloodstream so that it might be kept in circulation.

This unexpected finding had implications that went beyond the question of insulin metabolism. The young investigators quickly characterized the large plasma protein that was binding radioactive insulin in the blood of insulin-treated subjects. It was a globulin, a large protein like an antibody molecule, which clearly appeared in response to injections of insulin. Yalow and Berson concluded that treatment with insulin injections immunized patients so that they developed insulin-binding antibodies. In other words, the injection of beef or pork insulin called forth the large sheltering molecules that would take the little insulin molecules into their midst and keep them in the bloodstream. It is this property of taking insulin up into the larger protein (the globulin that has appeared in response to injections of insulin) that makes that globulin an insulin-binding antibody.

Berson and Yalow prepared their data for publication in the meticulous style that was already their hallmark. But the most prestigious scientific journals rejected their work, first *Science*, and then the *Journal of Clinical Investigation*. They were incensed, not simply because their paper was rejected, but because they were unable to convince the peer reviewers and editors with the logic of their position, and because the editors, in rejecting their insistence that the globulin was an antibody, accused them of "dogmatism" and characterized their writing as "incoherent."[71] The scientific world was not ready to accept the fact that a molecule as small as insulin could stimulate the human body to produce antibodies. Letters flew back and forth, passions flared, and then, in the midst of the battle, they accepted a compromise to drop the term antibody in the title of the paper and call the protein an insulin-binding globulin. Nonetheless, at that moment, they moved past the argument because they were on to something even more exciting.

Yalow and Berson had observed that radioactive insulin could be displaced from the sites on the antibody molecule that bind the radioactive insulin by the native insulin in the patient's blood samples. In other words, the radioactive insulin simply competed with the native insulin for a special place on the big antibody molecule, like when kids try to sit on the same chair. When they determined the precise amount of radioactive insulin that was pushed off the antibody molecules by a known volume of a patient's blood and compared that with the displacement produced by known concentrations in standard insulin preparations used for treating diabetics, they were measuring insulin concentrations in human blood.

They called the method radioimmunoassay: *radio* because the antigen (a substance foreign to the body that initiates an immune response, e.g., antibody production), in this case insulin, was labeled with a *radio*active isotope of iodine (I–131); *immuno* because the antigen (insulin) was binding to an antibody, so at the heart of the method was an *immuno*logic reaction between an antigen and its antibody; and *assay* because they were measuring something. Since almost any molecule can be an antigen, meaning that when injected into an animal in the proper circumstances almost any substance can be antigenic and stimulate the production of finely tuned antibody molecules, which are absolutely specific in recognizing only that stimulating antigen, Yalow and Berson understood that their method would have many applications.

RIA, as it soon became known, can be likened to the children's game of musical chairs, if we may extend this metaphor. In an RIA, the radiolabled antigen competes for antibody binding with the native antigen in a patient's blood sample. The antibody molecules, with their special ability to fit and bind with specific antigen molecules, are the chairs. The antigen molecules are the kids. We can see the kids only if they are wearing party hats. The radioactively labeled antigen (e.g., insulin-I–131, insulin molecules that have been labeled with I–131, a radioactive isotope of iodine) are the kids wearing the party hats. The molecules of a patient's own insulin in a small volume of blood plasma are the kids without hats.

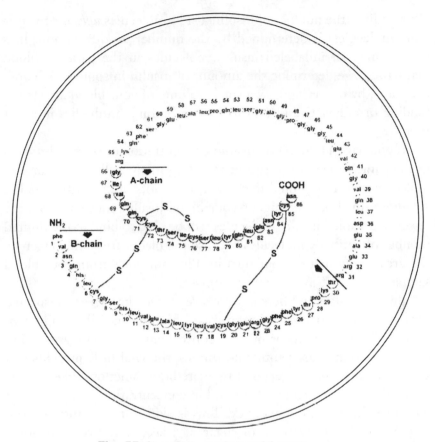

Fig. 33. Insulin's amino acid sequence.

In playing the RIA game, we can use as many chairs as we like because we can adjust the antibody concentration in the assay. We can have as many kids with hats as we want because we can add as much labeled insulin as we like. What we want to find out is how many kids without hats are in a small volume of the patient's blood plasma. We start the game by putting the three ingredients—the antibody molecules (chairs), the labeled insulin (kids with hats), and the small volume of a patient's plasma (containing an unknown number of kids without hats)—into a test tube with a buffer solution (a room with music).

When we stop the music, the number of kids with hats who are left standing (the number of insulin-I–131 molecules *unbound* to antibody molecules) is determined by the number of kids without hats (the number of unlabeled insulin molecules in the patient's blood plasma). So we determine the amount of insulin in someone's blood, say yours, by observing how a small volume of your blood affects the binding of radioactive insulin to insulin-binding antibodies in a test tube.

If you have a high blood insulin concentration (lots of hatless kids in the game), then at the end of the assay many radiolabeled insulin molecules will fail to bind to the antibody (lots of kids with hats left standing). In short, the higher your blood insulin concentration, the fewer radiolabled insulin molecules will be able to bind to the limited number of antibody molecules because the places are simply taken by the greater number of your own insulin molecules from your blood sample.

By 1955 Yalow and Berson were already well known and respected for their quantitative work in the areas of iodine and albumin metabolism (albumin is a major protein constituent of blood plasma). But the ability to measure peptide hormones, the vital little proteins that act as chemical messengers and regulate body functions, like insulin, some of which circulate in the blood in concentrations as low as 10^{-12} moles per liter, was quite a trick. This is akin to measuring the concentration of one teaspoon of sugar in a lake sixty-two miles long, sixty-two miles wide, and thirty feet deep. They quickly realized that they could apply the method to a vast array of endogenous (produced by the body) and exogenous (introduced from outside the body) substances. This was because antigenicity, the ability to elicit an antibody response, as they had shown for insulin, was a far more general characteristic of molecules than had previously been thought. The commercial possibilities for RIA were enormous, and indeed RIAs have made fortunes for commercial laboratories and producers of assay kits.

"We never thought of patenting RIA," Yalow told me, looking down her nose as though a dead fish had been placed before her. "Of course, others suggested this to us, but patents are about keeping things away from people for the purpose of making money. We wanted others to be able to use RIA. Now some people assume that I'm sorry, but I'm not. Anyway, we had no time for such nonsense."[72]

The antibody is the heart of any immunoassay because, like a deft and educated hand, it must grasp only one specific molecule, its antigen, the very molecule that called it forth from the lymphocytes of the immune system. In the assay, that same antigen must be picked out by the antibody from any number of other, even similar, molecules it encounters. In other words, because almost any chemicals can stimulate antibody production, Berson and Yalow created a tool that could be adapted for work on almost any biological problem, be it a question relating to insulin metabolism in diabetics, the reason why some people get hepatitis after a blood transfusion, or the determination of whether a patient has too much or too little of a specific medication in his blood.

Yalow and Berson never had a large or well-equipped laboratory, but their rapid application of RIA to a variety of scientific issues captured the imagination of the scientific community. Many brilliant young physician-scientists flocked to their mom-and-pop operation from the early 1950s through the late 1960s. Yalow and Berson first transformed the science and practice of endocrinology (the study of hormone action) by developing assays and studying insulin, growth hormone, adrenocorticotrophic hormone (ACTH), and parathyroid hormone (PTH), among others. They then showed the way to new vistas in hematology, gastroenterology, virology, and pharmacology, while teaching the rest of the scientific world their methods. The result was exponential progress in virtually every aspect of medicine and biological science—every clinical, research, and commercial medical laboratory in the world was using their method, every pint of blood was checked for contamination with it—and chances are that your blood has been analyzed by RIA.[73]

42.

The Discovery of Viruses

Q. How could we see what could not be seen?
A. By standing on the shoulders of giants.

Leeuwenhoek was the first to see protozoa, fungi, bacteria, sperm, and blood cells (see advance 4). But more than that, he taught us that there is a diverse and vibrant world that we can't see unaided and that its inhabitants can cause disease. Louis Pasteur (advance 15) and Robert Koch (advance 44) developed the germ theory and invented methods of isolating, growing, and proving that specific germs cause specific diseases. However, their work encountered a major problem: there were infectious diseases for which no infectious agent could be found, although Koch's postulates could otherwise be fulfilled. And so, for example, in the 1880s Pasteur was developing a vaccine for rabies without understanding its cause.

Little by little evidence accumulated, and the unseen began to be conceptualized. In 1880 the German scientist Adolf Mayer was working with a blight on tobacco leaves called tobacco mosaic disease. He found that he could transmit the disease from leaf to leaf and plant to plant using extracts of infected leaves. But he was unable to isolate a bacterial or fungal agent. He considered that the agent might be a soluble substance, like a toxin. But he gave up this line of thinking in favor of the idea that it was a new type of bacterium. In 1892 the

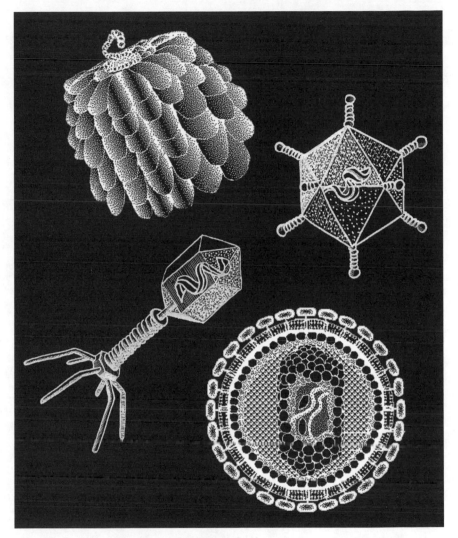

Fig. 34. A few viruses are harmful, but most have fascinating shapes like those shown above.

Russian researcher Dimitri Ivanofsky published his finding that the tobacco mosaic agent could pass through a porcelain filter known to block the passage of bacteria, but he, too, could not isolate the organism. Ivanofsky first questioned the integrity of his filter, and then, in 1903, he concluded that the agent was a toxin and not a reproducing organism. The Dutch researcher Martinus Beijerinck showed that the filtered infectious agent was not a toxin because it could grow and reproduce in the cells of the tobacco plant. In 1898 he published that this was a new infectious agent, a *contagium vivum fluid* (contagious living liquid). The term that stuck was "filterable agent," and there followed a decades-long controversy as to whether filterable agents were liquids or particles.

In 1917 Felix d'Herelle, a Canadian, discovered that filterable agents could infect bacteria, causing holes or plaques in the colonies they formed when growing on solid media. He called these agents bacteriophages. Each plaque developed from a single bacteriophage. This observation led to the "plaque assay," the first method for quantitating viruses. In 1935 Wendell Stanley crystalized tobacco mosaic virus, proving that viruses have a regular shape. In 1938 tobacco mosaic virus was seen through the electron microscope.

Viruses infect all life-forms on Earth. They consist of genetic material—either deoxyribonucleic acid (DNA) or ribonucleic acid (RNA)—surrounded by a protein coat, called a *capsid*, with or without an outer lipid envelope. They are from twenty to one hundred times smaller than bacteria, and they are measured in nanometers (10^{-9} meters). The largest viruses, the poxviruses, are about 450 nanometers (0.000014 inches) in length, and the smallest, polio viruses, are about 30 nanometers (0.000001 inches). Viruses cannot reproduce outside of a living cell. They transmit their genetic material from one cell to another in the process of replication, and, in doing so, they often damage or kill the cell they have infected, causing diseases like the common cold, flu, yellow fever, ebola, or AIDS, among many others. Some viruses cause cells to grow uncontrollably and trigger cancer. Many viruses infect cells but cause no symptoms or disease.

Perhaps most significantly, viruses are proving to be great instructors and facilitators. The fact that viruses are essentially bags of DNA or RNA means that by studying them we learn about normal cell biology and genetics. They are also being used as tools for the industrial and research production of proteins and for transporting genetic material into cell nuclei for gene therapy. Viruses can be engineered to kill selected cell populations, such as cancer cells. Genetically engineered insect viruses are being used as pesticides. And they can be made to kill selected human populations. And take note: attempts have been made to weaponize every class of infectious agent.

A controversial topic is the potential patenting, owning, producing, and distributing of these viral products. Remember—the genomes of some living organisms have already been patented.

"Those opposed to cloning and to patenting living things say the patent is a further sign that human life is being turned into a commodity."[74]

This is something to seriously ponder.

43.

Retroviruses and Oncogenes

A mong the great and immediate effects of the discovery of DNA's structure was the attraction of the best minds into what became known as molecular biology and molecular medicine. These people understood nucleic acids (DNA and RNA) in a visceral way, and they began to play with them like children with Play-Doh. They tried to merge red dough with blue, RNA with DNA, viruses with animal genes, and, while understanding the few rules of the new biology, they were not intimidated by the power of their playthings. The result has been an exponential understanding of knowledge and insight that informs our appreciation of life, our ideas about normality and disease, and our ideas about our origin.

Two of these extraordinary people were Harold Varmus and Michael Bishop who, while working at the University of California at San Francisco Medical School with the Rous sarcoma virus (RSV), exposed the processes by which certain viruses (retroviruses) can hijack cellular genes and turn them into cancer-causing genes (oncogenes). Fortunately, such transformations are not common, but they teach us about the basic biology of viruses and our own human cells. The understanding of retroviruses and oncogenes is one of the greatest medical advances and already has extensive practical utility.

Another great scientist, Peyton Rous (1879–1970), was a research pathologist working at the Rockefeller Institute in New York. In 1909 a poultry farmer brought him a chicken with a malignant tumor. Rous removed the tumor, ground it up, made cell-free filtrates, and injected the filtrates into normal chickens. The normal chickens developed similar malignant tumors (sarcomas). Rous then isolated the cancer-causing virus from the filtrates. It was the first time that a

Fig. 35. Harold Varmus.

virus was shown to cause malignancy, which was a discovery of vast proportions. Decades later certain human malignancies were shown also to be caused by viruses.

In 1969 Varmus and Bishop made plans to begin studying the ways that genes of RNA tumor viruses might combine with chromosomes of mammalian host cells. Shortly before they started, Howard Temin and David Baltimore independently discovered the method: a startling enzyme, reverse transcriptase, which is found in virus particles. An enzyme is a protein that can facilitate a chemical reaction. Reverse transcriptase (see advance 72) can synthesize DNA from RNA, so that an RNA tumor virus (a retrovirus, like RSV) can make DNA from its RNA genomic template.

Varmus and Bishop then set out in the early 1970s to use reverse transcriptase to make DNA probes from RSV RNA in order to define the chicken genes that RSV interacts with to produce malignant disease. They were aware that in 1970 a mutant strain of RSV had been detected which lacked the ability to transform normal chicken cells into malignant cells. So the RNA structure of this mutant strain (tdRSV), they thought, must differ from RSV by lacking an RNA sequence that determines the structure of the transforming factor.

In short, they took RSV RNA (containing the transforming factor sequence) and transcribed it with reverse transcriptase to obtain RSV DNA. Then they added tdRSV RNA and hybridized the RNA to the DNA (in hybridization the corresponding RNA and DNA sequences bond together in a process called annealing). But the transforming factor sequences in the RSV DNA had no RNA to anneal to and so were left free. The scientists retrieved these free DNA sequences, which gave them a product that they called sarc DNA (sarc = sarcoma), or a sarc probe.

The team used their probe to discover that the RSV transforming gene is represented in the normal avian bird cellular genome (DNA) as a normal gene, which, in a slightly altered form, became part of a retroviral structure way back when RSV was evolving. Even more striking, they concluded that the avian progenitor of the viral oncogene has been conserved through at least 100 million years of avian evolution because it plays an essential role in the life of birds—that is, influencing normal control of cell growth and development. So the pathogenic viral oncogene was derived from a beneficial proto-oncogene, and very subtle differences in the amino acid sequences of the src-proteins they encode for are responsible for sarcogenesis instead of normal growth. In other words, the harmful oncogene evolved from an essential cell-growth control gene.

Using these tools and concepts, investigators continue to advance our understanding of the genetic basis of cancer, the transposition of DNA through RNA intermediates, the control of gene expression (when, where, and how much protein a gene produces), and the molecular evidence for evolution. Harold Varmus went on to become the director of the National Institutes of Health. In that position he was innovative, inspiring, and greatly respected by his colleagues. After many years of strengthening the NIH, he became the director of the Memorial Sloan-Kettering Cancer Center.

44.

Koch's Postulates

L ouis Pasteur (see advance 15) had a sometime colleague and rival in the monumental work of understanding the role of microorganisms and in establishing microbiology as a scientific discipline: Robert Koch (1843–1910), and his meticulous laboratory techniques for identifying bacteria and growing them in pure cultures led to his associating certain bacteria with specific diseases. The key to his contribution was precision and technical innovation. He developed several types of solid-culture media, for which he employed the petri dish. He formed a large talented team, for which he built an institute.

Koch, a German, set down his method for determining the bacterial cause of disease in 1879 and modified it in 1882 as a series of four postulates. These have guided physicians and microbiologists to this day. But they are not commandments, and they have been flexible enough to survive the discovery of viruses, bacteriophages, and other nonbacterial pathogens. In fact, they helped in their discovery. Their results have included the accurate identification of the causes of countless microbial diseases. In addition, Koch's methods facilitated the identification of bacterial and other toxins; the discovery of passive immunity; and the development of antitoxins for diphtheria, tetanus, cholera, plague, and snake bites.

Koch's postulates demand that to establish the microbial cause of a disease one must demonstrate that:

1. the organism is discoverable in every case of the disease;
2. once recovered from the body, the organism can be grown in pure cultures for several generations;
3. the disease can be reproduced in experimental animals through a pure culture removed by several generations from the organisms initially isolated; and
4. the organism can be isolated from the inoculated animal and recultured.

Koch worked out these postulates while engaged as a district medical officer in a busy general practice of medicine. His laboratory, where he worked with anthrax bacillus, was in a four-room flat that he shared with his wife. There he kept mice and his equipment, mainly a microscope, which his wife had bought. (I wonder what other contributions she made.) But his drive and accomplishments were right in character. As a five-year-old he had astounded his parents when he showed them that he had taught himself to read by studying their newspapers.

Koch discovered the tubercle bacillus (the cause of tuberculosis) in 1882. In 1883 he took a trip to Egypt and discovered vibrio cholera. His rules for containing cholera are still in use today. He prepared extracts of tubercle bacilli, which he called "new tubercullin" and "old tuberculin," and while their therapeutic value was, to put it mildly, disappointing, his tuberculin skin test, in which a small volume of old tubercullin is injected into the skin and observed for a reaction forty-eight hours later, is still commonly used. Chances are you've had it at least once.

Pasteur's great public successes with the dramatic anthrax and rabies vaccinations put Koch under immense pressure from the German press and government. He had become an icon of German culture and was expected to outshine Pasteur. German superiority was at stake. At the Tenth International Medical Conference in Berlin, Koch announced that he had found a cure for tuberculosis. It was his tuberculin.

Did he already know that it didn't work? What was probably unknown to him was that it still contained active virulent organisms. He administered the tuberculin to hundreds of people who then developed TB and died. In the midst of a public scandal, Koch fled Germany for Africa where he worked on tropical diseases. The German government dealt with this defeat by setting up the Institute for Infectious Disease, sometimes called the Koch Institute. But the scientists and doctors involved began arguing over royalties for discoveries, and the institute fell apart only to rise again, several times, under new names.

Koch returned to Germany and discovered that the bacilli that caused human and bovine tuberculosis are not identical. He presented this finding at the International Medical Congress on Tuberculosis in London in 1901. It caused much controversy and opposition because he suggested that it could serve as a vaccine. Not likely, thought many stunned listeners. Some said worse. But Koch was right, and BCG (a strain of bovine tubercle bacillus) is now used as a vaccine, especially by the British. It is of limited value however, since its protective function is weak.

It's sad that Koch's career ended in scandal. Was it hubris? We blame the intrusion of nonscientific influences on medicine. Whatever it was, one of Koch's greatest lessons in this cautionary tale is the intrusion of political expediency into the scientific agenda. This continues as a current problem.

As the *New York Times* reported on March 1, 2005, more than seven hundred fifty scientists signed a petition sent to the National Institutes of Health complaining about the redirection of "tens of millions of dollars in federal research money since 2001 away from pathogens that cause major public health problems to obscure germs the government fears might be used in a bioterrorist attack."[75] Included in that group were Nobel Prize winners and a biologist scheduled to receive the National Medal of Science from President Bush later that month.

45.

Victory at C and the Concept of Vitamins

T
he idea that disease can be caused by the absence of something seems simple enough, but it took millennia, and developing chemical sophistication, for an understanding of vitamins to emerge. The diseases were known from antiquity—scurvy, beriberi, pellagra, pernicious anemia, rickets—and, in some cases, the salutary effects of certain foods were recognized. Still, the notion that in each case the lack of minute amounts of a specific chemical brought about ailments eluded scientists until the twentieth century. Then, with the isolation of vitamins, the determination of their chemical structures, and their synthesis, common disorders could be treated and prevented. It has been estimated that from 1600 to 1800 one million European sailors died of scurvy alone.[76]

A vitamin is a chemical, obtained from a food, which is required for life and health. A chemical may be a vitamin for some species, but not for others, since the others may be able to synthesize it within their bodies. Early in our appreciation of vitamins, they were called

194

"vital amines" because it was believed that they were all amines, or nitrogen-containing compounds derived from ammonia (NH_3). When it was found that they were not all amines, "vitamine" had been so accepted that the *e* was dropped, and the term survived. There are thirteen vitamins in all, divided into the four fat-soluble (A, D, E, and K) and the nine water-soluble (eight B vitamins and vitamin C) ones. The fat-soluble vitamins can be stored in the body and do not need to be ingested every day. Because they can be stored, it is possible to store too much and thus become poisoned. The water-soluble vitamins cannot be stored, with the exceptions of B12 and folic acid, and must be consumed frequently for optimal health. However, these water-soluble vitamins can be taken in larger amounts without toxicity, because they are easily eliminated. So an essential vitamin is a substance that the body cannot synthesize on its own yet is necessary for life. By definition, it is necessary to obtain all vitamins from outside the body. If a molecule can be synthesized in the body, it is not a vitamin. The exceptions to this rule are vitamin D, which can be synthesized in the skin, but only when exposed to sunlight, and niacin (B3), which can be synthesized in the liver in amounts too small to prevent deficiency disease.

Scurvy (vitamin C deficiency) was described in Egypt as early as 1500 BCE as occurring in the winter when fresh produce was unavailable. Hippocrates (see advance 3) described scurvy in the fifth century BCE as connected to bleeding gums, hemorrhaging, and death. Scurvy played a role in the Crusades and other historic events, but it was in the age of exploration, on those extended sea voyages, that scurvy earned its deadly reputation. Columbus's crew is said to have put some dying Portuguese seamen on an island only to find them all well upon their return. The island was rich in fruits, so they named it Curacao (cure.) A ship's company, with its dietary restrictions and uniformity, proved to be a laboratory for observing nutritional disease.

American Indians had a cure for scurvy. They drank a tea of pine bark and needles. Jacques Cartier, a French explorer, brought this cure back to France in 1536. He had lost twenty-five seamen before the natives taught him this remedy. But it was rejected by the medical profession. They were not prepared to learn from heathen savages.

Thus, many medical advances have been lost over the ages to narrow-mindedness, greed, prejudice, and pride.

In 1593, during a voyage to the South Pacific, Sir Richard Hawkins recommended the following treatment for scurvy: "That which I have seen most fruitful for this sicknesse, is sower [sour] oranges and lemmons."[77] In 1601 Captain James Lancaster captained one ship among a fleet that left port in late April and arrived at their destination in September of that year. The other ships had been so devastated by scurvy that Lancaster's men, all fit and healthy, had to help the other ships into the harbor. Lancaster sent a report to the admiralty explaining that he had brought on board bottles of lemon juice and every man took three spoonfuls each morning.

John Woodall's book, *The Surgeon's Mate*, published in 1636, stated with regard to scurvy, "The juyce of lemmons is a precious medicine and well tried; being sound and good. Let it have the chief place for it will deserve it. The use whereof is: It is to be taken each morning two or three teaspoonfuls, and fast after it two hours. Some chirurgeons also give of the juyce daily to the men in health as preservative."[78]

In 1747 the Scottish physician James Lind, aboard the *Salisbury*, experimented with patients suffering from scurvy. He fed them carefully controlled diets and discovered that those who were given an orange and lemon combination recovered promptly. Nevertheless, it took until 1770 for the British navy to begin recommending that ships carry lime juice for all aboard. When the colonies revolted, the colonial army used fruit to keep their soldiers healthy; they used the Indian remedy of pine bark and needles when fruit was out of season. An antiscurvy (antiscorbutic) vitamin was postulated in 1911. It was the third (or "C") vitamin to be discovered. It was antiscurvy or ascorbutic, so it got the name "ascorbic acid." Albert Szent-Gyorgyi isolated the substance ($C_6H_8O_6$) in 1928.

But at the turn of the twentieth century, the germ theory of disease (see advance 15) was so entrenched in scientific thought that beriberi (thiamine [vitamin B1]—[$C_{12}H_{17}N_4OS$]Cl—deficiency) and pellagra (niacin—$C_6H_5NO_2$—deficiency) were believed to be caused by bacterial infections or toxins. And the stories of pellagra and beriberi are full of medical ineptitude and governmental indifference. They are not unlike today's Gulf War syndrome, which should be studied. Together they emphasize the importance of controlled experiments, the scientific method, and the responsibilities that can be traced back to Apollo and before. These responsibilities for compassion and honesty are as important as ever.

Henry James's admonition about writing is equally applicable to medicine. "The effort really to see, and really to represent," James wrote, "is no idle business in the face of the constant force that makes for muddlement."[79] Notice that there are many missing letters in the sequence from A to K, and missing numbers in the sequence between B1 and B12. These blanks turned out to be vitamins that never were, products of "the constant forces." The vitamin and nutritional supplements industry quickly grew very large in the latter half of the twentieth century. The mystique of vitamins as potent cures and their relative safety have provided happy hunting for frauds and charlatans. In addition to their legitimate uses, vitamins are sold as tonics and cures for all manner of disease.

46.

Carbolic Acid Wound Dressing

The operation was a success, but the patient died—this oxymoronic maxim is much of what you need to know about the business of surgery before Joseph Lister (1827–1912) hit the scene in 1865. The British surgeon, who would unknowingly lend his name to the antiseptic mouthwash Listerine, entered a medical profession in which approximately 50 percent of surgery patients died as a result of postoperative infection. And without discovering a single new drug, he almost revolutionized his profession and came to be known as the Father of Antiseptic Surgery.

It was also in 1865 that Louis Pasteur discovered that the transmission of airborne organisms caused the decay of bodily tissues, and Lister hypothesized that those same microbes were the likely source of wound sepsis. Even before the work of Pasteur on fermentation and putrefaction (advance 15), Lister had been convinced of the importance of scrupulous cleanliness and the usefulness of disinfectants in the operating room; and when, through Pasteur's researches, he realized that the formation of pus was due to bacteria, he proceeded to develop his antiseptic surgical methods. As Pasteur had been earlier, Lister was rebuked.

Nevertheless, in August 1865 Lister successfully treated the wound of an eleven-year-old boy who had sustained a compound fracture of his tibia with carbolic acid, a chemical being used extensively in the local sewer system. Two years later, he presented his findings to the British Medical Association, and surgery has never been the same since. Within five years, the mortality rate of surgical patients fell from 50 percent to 15 percent. At this early stage, however, Lister's approach was to coat the wound with a potent dose of carbolic acid (a coal tar derivative), forming an antiseptic crust of coagulated blood on the wound. Unfortunately, carbolic acid is extremely toxic and can severely damage the body's repair mechanisms.

To combat this phenomenon, Lister developed a number of antiseptic wound dressings designed to provide a disinfectant barrier between the wound and the surrounding air, while reducing to a minimum the carbolic acid entering the damaged tissue. He also introduced the sterilization of surgical instruments with heat and carbolic acid and encouraged surgeons to clean their hands frequently with a mild antiseptic soap during operations.

Lister's methods of asepsis are the foundation of modern surgical technique. From the design of the operating room (OR) suite, to the processes of scrubbing, to the OR garb and protocol, Lister's legacy lives on.

47.

Dentistry

My mouth is full of decayed teeth and my soul of decayed ambitions.

—James Joyce (1882–1941), letter to his brother, February 19, 1907

It is generally accepted that modern dentistry was pioneered, in the Western tradition, by a French surgeon named Pierre Fauchard (1678–1761), who practiced in Paris from about 1715. He brought the trade of "tooth drawing" to a profession by focusing on a scientific approach to the anatomy, physiology, and pathology of the teeth. In the process he changed the dentist from a flamboyant showman, stepping into the midst of suffering to extract the affliction with a dramatic tug, to a scholar-practitioner with all the inclinations of a medical scientist. He was concerned with the dissemination of dental knowledge and wrote *The Surgeon Dentist* (1728; trans. of 2nd ed., 1946), long a standard work on dentistry. Fauchard set the agenda by viewing extraction as a last resort, a failure rather than a triumph. Preservation of teeth and function was now the goal, and he introduced or refined a host of techniques including filing, drilling, filling, straightening, denture fabrication, and surgery of the gums and jaw.

In colonial America most dentistry was practiced by itinerants whose training was primarily gained through apprenticeship. With the development of medical schools, some dentistry was taught, and physicians would perform extractions and other dental procedures. But, as in Europe, professional dentistry grew as a subspecialty of surgery, and the notion that it was founded upon the practice of barbers, bloodletters, and mechanics relates to the parallel trade in tooth drawing which persisted out of necessity and still does among the millions who lack dental insurance. Any doctor who hasn't seen citizens with amateur dental work hasn't been around much or hasn't looked into many mouths.

Early American books on dentistry such as *A Treatise on the Management of the Teeth* by Benjamin James, MMSS (member of the Massachusetts Society of Surgeons), published in 1814, and *The Family Dentist*, by Josiah E. Flagg, MD, MMSS, published in 1822, were written by people with medical backgrounds. In 1829, eleven years before the first dental college in the world opened in Baltimore, Samuel Sheldon Fitch, MD, published *A System of Dental Surgery* (2nd ed., 1835).[80] This was the best text of its time in the United States, with an extensive and international bibliography. Fitch was a graduate of Jefferson Medical College of Philadelphia, and he included a chapter on disorders caused by diseased teeth including headaches and inflammation of the facial nerve, eyes, ears, and brain.

The first college in the world devoted exclusively to dentistry opened in Baltimore in 1840. So the splitting began, and, as with other specialization of fields, there have been benefits and losses. With the opening of more dental schools, the profession gradually became independent. Physicians today know very little about dentistry, though both fields have made great advances both independently and collaboratively. The development of anesthesia (advance 35), the management of infection, the central role of prevention with hygiene and water flouridation, the use of metals and other materials, the use of high-speed turbine drills in surgery, and the use of dental techniques to fabricate maxillofacial prosthetics to reconstruct the faces of cancer victims and for plastic surgery are only a few examples of the positive interaction between medicine and dentistry.

Still, we are a unified organism, and virtually all processes involve all parts of the body. Whether it is the effects of diabetes on the mouth's bones and gums or the relationship between periodontitis and valvular heart disease, we must take an inclusive, integrated approach. The life expectancy of teeth cannot be split from that of the rest of the body, and yet there are many people, even those with medical insurance, who have no dental coverage.

Suggested Reading

1. Milton B. Asbell, *Dentistry: A Historical Perspective* (Pittsburgh: Dorrance, 1988).

2. Samuel S. Fitch, *A System of Dental Surgery* (Philadelphia: Carey, Lea & Blanchard, 1835).

3. Arthur W. Lufkin, *A History of Dentistry* (Philadelphia: Lea & Febiger, 1938).

4. Malvin E. Ring, "Dentistry's Contributions to Medicine," *Journal of the Maryland State Dental Association* 34, no. 4 (Winter 1991).

48.

The Relaxing Factor That Led to Viagra

s a medical student I became fascinated with the work of my professor of pharmacology, Robert Furchgott, who was working with helical strips of rabbit aorta to study vasoactive substances (chemicals that contract or relax arteries) and their actions at membrane receptors. His personal involvement of going to the lab, watching and discussing experiments with technicians and postdoctoral students, and warmly welcoming and taking students seriously began a friendship that helped to shape a twenty-year period in which I was, almost exclusively, a laboratory investigator. Little did I then appreciate the momentous discovery that he was going to "accidentally" (his term) make with the system that was before my eyes. For me, the fascination was simply in the direct connection between the intellectual constructs of words like catecholamines, sodium nitrite, histamine, and acetylcholine (ACh), and the physical demonstration of their reality in the moving aorta. At that time receptors were only an idea, as was the effector-receptor interaction. It would be years before receptors would be characterized as proteins, with known amino acid sequences and three-dimensional configurations. But I could see the molecules struggling for dominion over Furchgott's strips of tissue.

The details of the various types of receptors and their roles in mediating contraction and relaxation of aortic smooth muscle cells are beyond the scope of this chapter. But it is important to understand that their characterization involved attaching the helical strips to a simple device that measured and recorded the time course and strength of contraction, and then bathing the strips in buffers containing various concentrations of vasoactive substances. The experimental protocols are very precise: the order of exposure and washout of chemicals is critical, as they establish an equilibrium with the bound and unbound molecules at the receptor. And the tissue preparation is very critical. Furchgott's team began using transverse rings of rabbit thoracic aorta, rather than the helical strips with which they had earlier observed the responses. But in May 1978 Furchgott's technician made a mistake. Early in the experiment, before blocking the alpha-adrenergic receptors (with dibenamine), he tested a chemical (carbacol) for its contracting activity before, rather than after (as called for in the protocol), washout of a previous test dose of norepinephrine. Under those conditions the response to carbacol was not contraction, but was partial relaxation of the norepinephrine-induced contraction. The unexpected relaxation of the aortic strips by norepinephrine (a "muscarinic" agonist) was exciting because it was in accord with the vasodilating action of muscarinic agonists in live animals.

It was then discovered that the relaxing factor resided in the innermost endothelial cells lining the aortic lumen and that this endothelial-derived relaxing factor (EDRF) was generally present in arteries from many species. And it was further found (also by way of "accident," which favors the prepared mind) that EDRF is caused by the simple chemical nitric oxide, one atom of nitrogen and one of oxygen (NO). It is now well established that NO is an important signaling molecule that acts in many tissues to regulate a diverse range of physiological processes. In 1998 Furchgott shared the 1998 Nobel Prize in Medicine or Physiology with Ferid Murad and Louis J. Ignarro for their work with NO.

The discovery that EDRF was in fact nitric oxide—a small gaseous molecule—has led to an explosion of interest in this field and resulted in many thousands of publications over the last decade. NO is involved in an astonishing variety of biological processes including neurotransmission, immune defense, and the regulation of cell death (apoptosis). It is a very short-lived molecule (with a half-life of a few seconds) produced from enzymes known as nitric oxide synthases (NOS).

Since it is such a small molecule, it diffuses rapidly across cell membranes and, depending on the conditions, is able to diffuse over relatively long distances. The biological effects of NO are mediated through its reaction with a number of targets such as heme groups, sulfhydryl groups, and iron and zinc clusters. Such a diverse range of potential targets for NO explains the large number of systems that utilize it as a regulatory molecule. As a consequence, abnormal NO synthesis is capable of affecting many important biological processes and has been implicated in a variety of diseases.

NO can be produced by a number of cells involved in immune responses. White blood cells can produce high concentrations of NO in order to kill target cells such as bacteria or tumor cells. NO has been shown to kill cells by disrupting enzymes involved in glucose metabolism, DNA synthesis, and mitochondrial function.

NO is involved in inflammatory processes that are critical to tissue maintenance and repair. It is important in regulating blood pressure and platelet function. It also acts as a neurotransmitter in the central and peripheral nervous systems, and is involved in regulating apoptosis in neurons.

As we have seen, NO was discovered through its ability to make blood vessels dilate by relaxing the vessels' smooth muscles. That is how it regulates blood pressure, and that is how it can help trigger erection of the penis, because the relaxation lets blood flow into the expandable sinuses of the organ. Viagra is designed to increase nitric oxide's effect.

49.

Advanced Imaging Techniques

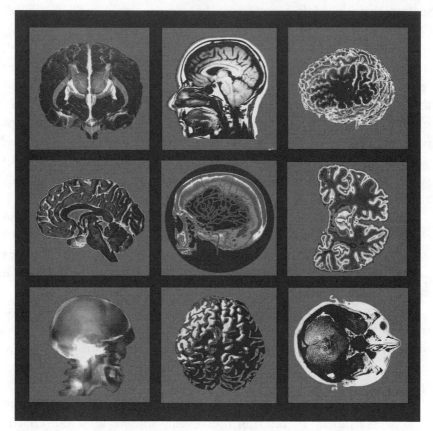

Fig. 36. The brain imaged by various techniques. *Left to right–Top Line*: 3D MRI brain scan, MRI brain scan, MEG scan. *Mid Line*–CT scan, angiogram, MRI scan. *Bottom Line*–X-ray, photograph, MRI scan.

Any mechanic who claims he can properly diagnose your dying car without opening the hood is probably a hoodwinker. In both medicine and mechanics a thorough visual inspection of the damaged or malfunctioning area is critical. Without that everything is guesswork.

Before advanced imaging techniques—CAT scans, PET scans, MRI scans, and ultrasound—were developed, all doctors had to go on were plain x-ray films (advance 11) and physical examination (advance 8) of patients. Advanced imaging technology opened the door to a new level of precision and success never before seen. For a doctor, working with these tools is like having x-ray vision, only better.

A computerized axial tomography scan (CAT scan) is a series of two-dimensional x-ray images transformed into three dimensions. A large donut-shaped tube is used to take x-rays of the same area from many differing angles (tomography). Those images are then put into a computer that combines the various angles in order to generate cross-sectional and three-dimensional images of the internal organs and structures of the body. This process was simultaneously and independently invented by Godfrey Newbold Hounsfield of EMI Laboratories, using gamma rays instead of x-rays, and by Allan Mcleod Cormack of Tufts University in 1972. They shared the 1979 Nobel Prize for their work. CAT scans are used to identify normal and abnormal formations in the body and assist surgeons by helping to accurately guide the placement of instruments or treatments.

Positron emission tomography (PET) is a nuclear medicine technique using a camera, which captures powerful images of the human body's function and reveals information of health and disease. Compounds normally existing in the body, like simple sugars, are labeled with radioactive tracers, which emit signals and are injected intravenously. The scanner records the signals that the tracer emits as it journeys through the body and as it collects in targeted organs. A powerful computer reassembles the signals into actual images, which then show biological maps of normal organ function and failure of organ systems in disease. PET scans are exceptionally dynamic in revealing functional activity. For example, you can actually see areas of the brain light up when the paitent moves, or even thinks, and you can see the difference in brain activity when a novice and a master musican listen to the same piece of music.

Ultrasound scanning, or sonography, is a method of obtaining images from inside the human body by using high-frequency sound waves. The reflected sound-wave echoes are recorded and displayed as a real-time visual image. The technology is based on the same principles involved in the sonar used by bats, ships at sea, and anglers with fish detectors. As the sound passes through the body, echoes are produced that can be used to identify how far away an object is, as well as its size, shape, and consistency. Because the images are captured in real time, ultrasound is a useful way of examining many of the body's internal organs. Ultrasound images can show movement of tissues and organs, and enable physicians to see blood flow and heart valve functions.

50.

Immunological Tolerance and Rejection

The meticulous surgical technique pioneered by French surgeon Alexis Carrel (1873–1948), especially in terms of his vascular work (he studied stitching with a lace maker), opened many doors. In fact, he successfully transplanted a variety of organs in animals, including kidneys and hearts. That is, the operations were successful, but in short order most of the animals died. The transplanted organ would be the sight of a strange inflammatory process leading to death. The host was intolerant, and the graft was rejected.

The understanding and manipulation of immunological tolerance and rejection is what has made successful organ transplantation possible. It has also allowed progress in the area of autoimmune diseases, like lupus, and in the treatment of malignant diseases.

Medicine, like music, literature, painting, architecture, and all other cultural expressions, goes through periods of focused insight. The dawn of these periods can be traced back to some inkling or intuition with a dramatic climax during which everything is seen through a new lens—all is excitement, everyone uses the new vocabulary— and then a newer discovery comes on the scene. But the old revelation is forever; the culture is enriched. So we have had the age of anatomy, of physiology, of microbiology, of surgery, of pharmacology, and now we are in the age of molecular biology and molecular medicine. When I entered medical school in the early 1960s, it was the age of immunology. The stars were Peter Brian Medawar (1915–1987) and Sir Frank Macfarlane Burnet, and they spoke about self-recognition and tolerance.

When tissues, like skin or blood vessels, or whole organs, like kidneys or livers, are transplanted between individuals of different genetic makeup, there is an immune reaction in which the host rejects the graft. The host recognizes that the proteins on the cell surfaces of the graft (tissue antigens) are not its own, they are foreign, and it mounts a complex cellular reaction that destroys the transplant. This cellular self-recognition develops early in life and is complete within a few months of birth. But a nonreactive, immunological tolerance can be caused by exposing animals to antigenic stimuli before they are old enough to undertake an immunological response. A human or an animal presented with foreign antigens during this window, from fetal life to shortly after birth, will then recognize the antigens as part of itself. Medawar described immunological tolerance as "a state of indifference or non-reactivity toward a substance that would normally be expected to excite an immunological response."[81]

A state of tolerance produced by this process is specific in that it will discriminate between one individual and another, thus an animal made tolerant of grafts from one individual will not accept grafts from a second individual (excepting an identical twin). But the tolerance will extend to all tissues from that donor.

The injection of white blood cells or lymphoid cells can confer tolerance of skin grafts, for example, and the same is true of grafts of other tissues such as of the thyroid, ovaries, kidneys, or adrenal glands. All of these various tissues differ in their antigenic makeup but not with regard to antigens that play a role in transplantation immunity.

The immune response that rejects foreign tissue is the work of mononuclear "lymphoid" cells such as lymphocytes, plasma cells, monocytes, and macrophages. These cells reside in lymphoid tissues such as the spleen and can be recruited to the site of the immune response through the bloodstream.

The host's immune response to a graft may lie anywhere across a broad spectrum of intensity, as immune tolerance is not an all-or-nothing phenomenon. All degrees of tolerance can be observed, from little or no response with permanent acceptance of the graft, to robust response that allows a graft to live just a bit longer than it would in a normal animal. An animal can be partially tolerant because all its immune-competent cells are almost completely debilitated or because, while most of them are completely out of action, a minority retain full possession of their powers.

During World War II Medawar was asked by the British Medical Research Council to study the problem of skin-graft rejection. His research, mostly carried out in mice, established much of the basis of transplant immunology. He made basic observations concerning the increasing intensity of the host's cellular response and increasing speed of rejection with repeated grafts from the same donor. These established the immunologic nature of the rejection process.

Even from this very brief discussion, I hope it is clear that disabling the cellular immunologic response of the host can be crucial to improving the viability of organ grafts. In 1951 Medawar used cortisone, which had just become available, to suppress the immune response and prolong graft survival. This is now accomplished using a variety of immunomodulatory drugs, but the fundamentals of the cellular mechanism of rejection of skin grafts provided the great advance that led to the successes we have today. And manipulation of tolerance and the creation of animals with cell-surface tissue antigens compatible with human self-recognition is actively being pursued.

Suggested Reading

1. F. M. Burnet, *The Clonal Selection Theory of Acquired Immunity* (Cambridge: Cambridge University Press, 1959).

2. J. Lederberg, "Genes and Antibodies," *Science* 129 (1959): 1649–53.

Solid Organ Transplantation

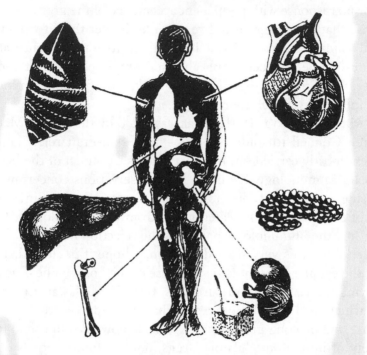

Fig. 37. Some of the solid organs that are now commonly transplanted. *Clockwise from upper right*: heart, pancreas, kidney, skin, bone, liver, lung.

O rgan transplantation can now give many years and even decades of nearly normal life to people who would otherwise suffer and die. But this great advance is a treatment and not a cure. It, in fact, demands other rigorous treatments requiring close cooperation between patient and transplant team. The transplanted organs require lifelong maintenance. But the rewards are high.

On February 4, 2005, there were 87,434 candidates on the United Network for Organ Sharing (UNOS) waiting list for donor organs. At that time the network reported that between January and October 2004, there had been 22,554 transplants performed, and there had been 11,766 donors. Therein lies the major problem in organ transplantation today: the disparity between the number of organs needed and the number available.

A major step in dealing with this problem was the formation and empowerment of UNOS. On its Web site you can find a great deal of useful information, including the following outline of its purpose, policies, and history:

> UNOS follows a collaborative policy development process for administering the Organ Procurement and Transplantation Network (OPTN) and handles the development, monitoring, enforcement and modification of the policies that govern the allocation, procurement and transportation of human organs.
>
> This process encourages participation by the public and all areas of the transplant community, including the government. It also promotes equity among patients waiting for organs and allows for ongoing and timely modification to reflect current science and medical practice.
>
> To become binding under the authority of federal regulation, the U.S. Department of Health and Human Services (HHS) must review and approve all policies and bylaws. . . .
>
> UNOS was awarded the very first OPTN contract on September 30, 1986, and has continued to administer the OPTN under contract with the Health Resources and Services Administration of the U.S. Department of Health and Human Services (HHS) for more than 16 years and four successive contract renewals. . . .

*　　*　　*

Timeline of Key Events in U.S. Transplantation and UNOS History

1954   First successful kidney transplant performed.

1966   First simultaneous kidney/pancreas transplant performed.

1967   First successful liver transplant performed.

1968   First successful isolated pancreas transplant performed.

            First successful heart transplant performed.

1968   The Southeast Organ Procurement Foundation (SEOPF) is formed as a membership and scientific organization for transplant professionals.

1977   SEOPF implements the first computer-based organ matching system, dubbed the "United Network for Organ Sharing."

1981   First successful heart-lung transplant performed.

1982   SEOPF establishes the Kidney Center, the predecessor of the UNOS Organ Center, for round-the-clock assistance in placing donated organs.

1983   First successful single-lung transplant performed.

            Cyclosporine introduced.*

1984   National Organ Transplant Act (NOTA) passed.**

1984   The United Network for Organ Sharing (UNOS)

separates from SEOPF and is incorporated as a non-profit member organization.

1986   First successful double-lung transplant performed.

1986   UNOS receives the initial federal contract to operate the Organ Procurement and Transplantation Network (OPTN).

1987   First successful intestinal transplant performed.

1988   First split-liver transplant performed.

1989   First successful living donor liver transplant performed.

1990   First successful living donor lung transplant performed.

1992   UNOS prepares first-ever comprehensive report on transplant survival rates for all active U.S. transplant centers.

UNOS helps found the Coalition on Donation to build public support for organ donation.

1995   UNOS launches its first Web site for all users with an interest in transplantation.

1998   First successful adult-to-adult living donor liver transplant performed.

1999   UNOS launches UNetsm, a secure, Internet-based transplant information database system for all organ matching and management of transplant data.

2000   U.S. Department of Health and Human Services pub-
            lishes Final Rule (federal regulation) for the operation
            of the OPTN.

2001   For the first time, the total of living organ donors for
            the year (6,528) exceeds the number of deceased organ
            donors (6,081).

---

*Cyclosporine was the first of a number of drugs that effectively treat organ rejection by suppressing the human immune system.

**The National Organ Transplant Act (P.L. 98-507) established the framework for a national system of organ transplantation.[82]

The UNOS timeline shown above is a helpful guide, although we dis-agree with the claim that cyclosporine was the very first drug that effectively treated organ rejection by suppressing the human immune system (see advances 50 and 66).

To date there have been about 333,000 organ transplantations in the United States, with an average of just below 25,000 per year since 2000. Data are available regarding race, gender, citizenship, and other demographics of the donors and recipients. Immunosuppressive treat-ments are improving, and the survival rates are increasing. In general, the one-, three-, and five-year survival rates for major transplanted organs (kidney, liver, heart, pancreas) are about 90, 80, and 70 per-cent, respectively. The goal is as it should be: to provide equal access for all people who need organ transplants, and UNOS does an excel-lent job. But there are real problems in achieving the goal, which are similar to those plaguing all aspects of our society. When you factor in the "business" of organ transplantation, the needs of the patients often get lost.

A search of the Internet can quickly reveal people arguing for organ auctions, dozens of ways to do end runs around UNOS, and stories of poor villages in India where many people have nephrectomy scars because they have sold a kidney to traveling harvesters.[83]

Kidney Transplantation

Almost everyone is born with two kidneys. Nevertheless, we need only one good one, and that redundancy of functional capacity is built into most organs. But the packaging of the functional units, which in the kidneys are called nephrons, into two essentially identical organs is especially convenient for donation and transplantation of kidneys. Add to that the fact that some very common diseases, like diabetes and high blood pressure, often cause chronic, end-stage kidney failure, and it is easy to see why 56,490 of about 87,000 people waiting for donor organs are waiting for kidneys. In the year ending on June 30, 2004, there were 15,523 kidney transplants in the United States, of which 8,985 were from deceased donors and 6,538 were from living donors.[84]

The best donor, of course, is the recipient's identical twin, and the first successful kidney transplant was done between twins in Boston in 1954. This eliminates the problem of rejection and the need for ongoing immune suppression.

A living related (nontwin) donor is more frequently available than a twin, however. There is a better chance of a good match than with a nonrelated donor. Receiving a kidney from a relative also avoids what could be a wait of years for a cadaver organ, which also might be damaged.

And then there is the real possibility of a spare-parts clone. In a free market economy where healthcare has often become a commodity and organs are bought and sold, it's only a matter of time. Maybe it has already happened.

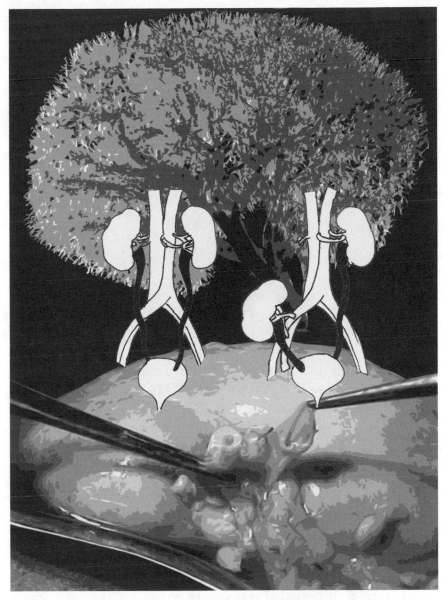

Fig. 38. *Top*—vasculature of the kidney. *Bottom*—donor kidney.
Center left is the normal anatomical arrangement of the kidney
and associated vessels and organs. *Center right* is the most
common posttransplant anatomy with the donor kidney attached
to the blood supply below the aortic bifurcation.

53.

Liver Transplantation

L iver transplantation is more complicated than kidney transplantation. The organ is so large, and the vasculature (arteries and veins) and bile ducts so complex and intimately related to the processes that cause end-stage liver failure, that the diseased host's liver must be removed. In contrast, diseased kidneys can often be left in place with the donor kidney put into the pelvis. This may shorten and simplify the surgery. And while the kidney-failure patient can be maintained rather well for an indefinite period by hemodialysis, there is no such successful remedy for liver failure. Thus, timing is more critical, and, as time slips by, the patient's ability to withstand the difficult operation diminishes. A greater proportion of liver patients die while waiting for a donor organ, as well as during the surgery. But once the organs have been successfully transplanted, the patient's survival rates are similar to those of kidney recipients.

The most common cause of liver failure leading to transplant in adults is called cirrhosis and may be due to alcohol damage or hepatitis viruses B and C. Hepatitis A occasionally causes acute liver failure and even more rarely death. But the vast majority of cases are self-limited, and, with time, the recovery is complete. In children, the most frequent reason for a transplant is a congenital developmental anomaly called biliary atresia, where babies are born with no bile ducts or very deficient bile duct development. Especially in children, there must be a rough size fit of the donor organ with the recipient's.

The liver has a marvelous ability to regenerate. When a portion is removed, even a very substantial amount, the organ continues to function, and, in short order, it increases in size and functional capacity. This has made it possible to have live liver donors.

*　　*　　*

Angela M. was an early liver transplant recipient. Her story is like many others, both typical and unique.

When I first met Angela, she was a frightened thirteen-year-old with a shy smile, but she was able to ask the hard questions: "Why did my eyes turn yellow?" "Am I very sick?" "Can I get better?"

Angela had brought her yellow eyes to the attention of the school nurse. The young student told of how her eyes had become a little yellow several months earlier and then had turned white again after about a week. This time, she explained, they were getting deeper yellow, and she thought she was sick. I did my best to answer her questions: I thought she had hepatitis. I wasn't sure how sick she was. Yes, I thought she could get better. I think she knew that she was quite sick, and she was waiting for the words that she could get better. It has been twenty years of growth and pain, but she's better now.

Angela is a trusting young woman and from the very first, from the time I said I thought she could get better, she had questions, but she fully cooperated and participated in her own care. One week after her first visit she returned, and I explained that we needed to do a liver biopsy. She asked a lot of questions and then said that she wanted it done. About a month later, when we had all the blood tests and we had studied her liver biopsy, I called her back. She was just a young girl with her mother, and I tried to explain her illness. She had autoimmune chronic hepatitis, and we don't know much about it except that it progresses, though we can delay the progress with treatment. She was happy to know that she might be feeling better soon. She was getting very tired by lunch time, and she didn't like her yellow eyes. She started taking steroids, and, in a short time, she improved. We were both thrilled by her improvement and became good friends. But Angela's liver biopsy looked pretty bad, and I began to think about liver transplantation.

At that time the results of liver transplantation were good, but not great, and there was no liver transplant center in New York State, so she would have to go to Pittsburgh, where Thomas Starzel had pioneered the procedure at the University of Pittsburgh. We decided to keep her going with medicines as long as possible. She finished high school. Sometimes she would get a little jaundiced and we would adjust the dose of medications. She grew tall and beautiful, and the results of liver transplantation kept improving. When she entered college, her liver biopsies began looking worse. She was now in her early twenties, and we started to discuss transplantation in earnest. But she felt good, got married, and had a beautiful little girl.

Angela would always bring her daughter when she came to see me. When the baby was almost two, Angela's legs began to swell, but we added diuretics and this, too, was controlled. Then Mount Sinai Hospital started doing liver transplantation in New York. Angela was still a strong young woman. Her daughter was talking now, and Angela had an almost 100 percent chance of surviving the operation. It was time.

Thank goodness for Medicaid. Angela worked but she could never have paid for the transplantations. Her first new liver lasted about two years before rejection forced a second transplant. The UNOS system of allocating precious organs in an equitable fashion and financial support for the procedures are triumphs of our culture. We must take pride in such achievements and extend them. Angela is doing very well. She is years out from her second transplant. She nonetheless speaks about how organ donation requires trust in the system and species pride and solidarity. She fears that this system may be threatened by human abuse and exploitation.

Resources

1) American Liver Foundation
   75 Maiden Lane, Suite 603
   New York, NY 10038
   Phone: 1-800-GO-LIVER (465-4837)
   E-mail: info@liverfoundation.org
   Internet: http://www.liverfoundation.org

2) Hepatitis Foundation International (HFI)
   504 Blick Drive
   Silver Spring, MD 20904-2901
   Phone: 1-800-891-0707 or (301) 622-4200
   Fax: (301) 622-4702
   E-mail: hepfi@hepfi.org
   Internet: http://www.hepfi.org

3) United Network for Organ Sharing (UNOS)
   P.O. Box 2484
   Richmond, VA 23218
   Phone: 1-888-894-6361 or (804) 782-4800
   Internet: http://www.unos.org

Heart Transplantation

Your heart is about the size of your fist. Normally, it beats about one hundred thousand times each day, pumping two thousand gallons of blood. But disease in the arteries that feed the heart, the strain of high blood pressure, alcoholic damage to the muscle, viral infection, or even unknown processes can cause the heart to fail. In children, congenital heart disease is the most frequent cause of heart failure. When the heart fails, it can't pump enough blood to sustain the body's functions. The symptoms are quite understandable from William Harvey's description of the heart as a pump in a closed circulation (advance 10). There are effects behind and forward of the failing pump: fluid may accumulate in the lungs causing shortness of breath (dyspnea); feet and legs may swell with fluid (edema); poorly fed muscles may cause fatigue; confusion may set in from a poorly perfused brain; and the patient may develop liver or kidney dysfunction. Drugs can help to improve the heart's performance: inotropic agents stimulate the heart muscle's contraction, antihypertensive agents lower the blood pressure and thus reduce the resistance to blood flow (after-load on the pump), and diuretics reduce the volume

224

of blood returning to the heart (preload). All these classes of drugs can work wonders and represent great advances, but for some patients the heart has lost too much function to work even with the drugs, so it must be replaced.

Some of the great technical advances of modern surgery have made heart transplantation possible: the heart-lung machine, cardiac hypothermia, the mechanical heart, methods to stop and restart the heart, among others. And among the great pioneering surgeons were C. Walton Lillehigh and Norman Shumway, and their colleagues in the United States. But the first human heart transplant was performed in South Africa in December 1967. The surgeon, a former student of Lillehigh, was catapulted to fame within a few days. The patient, a dentist with terminal heart failure, died in eighteen days. But the cat was out of the bag, and the race was on to be the first surgeon in Brooklyn, in Houston, at St. Elsewhere, to perform a heart transplant. The "ethical" issues revolving around heart donation, like the definition of death, the seat of the soul, black hearts and white hearts, all faded away.

To date there have been 36,435 heart transplants in the United States, with an average of just over 2,000 a year since 1990.

55.

Bone Marrow Transplantation

As part of my orientation to the Laminar Air Flow rooms, I was given a tour of the unit . . . four plastic sheetlined cubicles, end to end, each with a window on one side and all facing a long busy nursing station on the other. The windows looked north over a cityscape I could get used to and perhaps even get some nice photos. . . . I was given a piece of paper entitled "Entry Into The LAF Unit." It included suggested items I could bring with me: "Pictures of family, friends, etc., radio, comforter, a favorite stuffed animal!, socks, slippers, pillow, pajamas, a baby soft toothbrush, a hat, cap or scarf, playing cards, books, VCR tapes, games you enjoy playing or hobbies you enjoy." Gee, this didn't sound so bad. . . .

I had been home for five days since a rehospitalization for the treatment of an infection. I was extremely weak and would have done anything to get a good night's sleep. Worse, was the nausea and vomiting every time I looked at food or even thought about it. With few options, [my husband] did well at trying to vary and enhance the preparations of Jell-O, rice pudding, cereal, toast, noodles, soft vegetables etc., but little helped. There was nothing appetizing about any of it. Naturally I was supposed to eat. No more IV's to depend on. I weighed 89 pounds and avoided mirrors. The medicine wasn't working! [My doctor] was making a house call that night. He had mentioned trying Marinol pills and I was excited by the idea of getting a prescription for marijuana.

—Fran Epstein, leukemia survivor, now eleven years post–bone marrow transplant[85] one marrow is the spongy red tissue at the center of bones. The marrow in the sternum (breastbone), skull, hips, ribs, and spine contains stem cells that are the source of the body's blood cells. These blood cells include leukocytes (white blood cells), which fight infection; erythrocytes (red blood cells), which carry oxygen to and remove waste products from organs and tissues; and platelets, cell fragments that enable the blood to clot. Each cell type comes from a different stem cell line. Dysfunction of these adult stem cells causes abnormalities in the number and function of the cells they produce. The resulting diseases, including the leukemias (in which excessive, frequently immature, white cells accumulate in the blood and tissues), aplastic anemia (in which blood cell counts are very low), immune deficiency disorders, lymphomas, and multiple myeloma, can now be treated, and often cured, by bone marrow transplantation (BMT). In addition, BMT can be used to rescue patients in whom aggressive chemotherapy for breast, ovarian, or other cancers has caused bone marrow destruction.

In the treatment of leukemia, lymphoma, multiple myeloma, and the other malignant bone marrow disorders, large doses of chemotherapy and/or radiation are required to destroy the abnormal marrow stem cells and diseased blood cells. These treatments often destroy normal cells found in the bone marrow as well, just as the aggressive chemotherapy used to treat other cancers often does. BMT enables more aggressive chemotherapy and/or radiation by allowing replacement of the diseased or damaged bone marrow after the aggressive treatment. In this way, a transplant can increase the likelihood of a cure, or at least prolong the period of disease-free survival for many patients.

In any BMT, the patient's diseased bone marrow is destroyed and healthy marrow is infused into the patient's bloodstream as in a blood transfusion (see advance 36). The healthy bone marrow migrates to the cavities of the large bones, engrafts, and begins producing normal blood cells. The source of the healthy marrow used in the BMT may be from the bones of a living (related or unrelated) donor with a good tissue antigen match (see advances 36, 50, and 51), from the patient her/himself, or from adult stem cells filtered from the blood of a donor or filtered from umbilical cord blood after the birth of a baby.

If bone marrow from a donor is used, the transplant is called an "allogeneic" BMT, or if the donor is an identical twin, "syngeneic" BMT. In an allogeneic BMT, the new bone marrow infused into the patient must match the genetic makeup of the patient's own marrow as perfectly as possible. If the donor's bone marrow is not a good match, it will perceive the patient's body as foreign tissue to be rejected. The resulting graft-versus-host disease (GVHD) can be fatal. On the other hand, the patient's immune system may destroy the new bone marrow. This is called graft rejection.

There is a 35 percent chance that a patient will have a sibling whose bone marrow is a near-perfect match. If the patient has no matched sibling, a donor may be located in one of the international bone marrow donor registries. Sometimes a mismatched or "autologous" (the patient's own marrow is removed and replaced after treatment) transplant may be considered.

In an autologous BMT, the disease involving the bone marrow is in remission, or the condition being treated does not involve the bone marrow (e.g., breast cancer, ovarian cancer, Hodgkin's disease, non-Hodgkin's lymphoma, and brain tumors). Here, bone marrow is extracted from the patient and may be "purged" to remove any lingering malignant cells (in the case of a patient in remission). The harvested bone marrow, usually about one to two quarts of thick red liquid, can be frozen (cryopreserved) before being returned to the patient.

In allogeneic BMTs, the bone marrow may be treated to remove "T-cells" (T-cell depletion). T-cells are involved with cellular immune responses. They are thymus gland–derived immune-competent lymphocytes, and their removal reduces the risk of graft-versus-host disease. The donated marrow will then be transferred directly to the patient's room for infusion. The donor will experience some discomfort at the site of the harvest, but their marrow (20 percent of the their total bone marrow) will be fully restored in a few weeks and the discomfort will disappear far earlier than that.

BMT is, perhaps more than other transplants, a physically and emotionally demanding treatment.[86] But the rewards are fantastic, and as virtually every patient has told me, even before the process was over, it is well worth the hardships. The month immediately following transplant is the most difficult. The high-dose chemotherapy and/or radiation given to the patient before the BMT will have destroyed the patient's bone marrow, nearly abolishing the body's immune system. As the patient waits for the transplanted bone marrow to migrate to the cavities of the large bones, establish itself ("engraft"), and begin to produce normal blood cells, she will be very vulnerable to infection and bleeding. These patients receive complex antibiotic regimens, as well as blood and platelet transfusions. Allogeneic graft recipients must take additional medications to prevent and control graft-versus-host disease. Extraordinary precautions must be taken to minimize the patient's exposure to infection and minor trauma. The patient must live in virtual reverse isolation. In other words, the patient is isolated not to protect healthy people from disease but rather to protect the vulnerable patient from dangers transported by healthy visitors.

Visitors should be kept to a minimum. All visitors and hospital personnel should wash their hands with antiseptic soap and wear protective gowns, gloves, and surgical masks while in the patient's room. The room will need to be specially equipped with air filters. Fresh fruits, vegetables, plants, and cut flowers will be prohibited because they often carry fungi and bacteria. When leaving the room, the patient will have to wear a mask, gown, and gloves. Daily blood tests are done to determine whether engraftment has occurred and to monitor organ function. Finally, as the transplanted bone marrow engrafts and begins producing normal blood cells, the patient will gradually be taken off the antibiotics, and blood and platelet transfusions will generally no longer be required. The typical hospital stay for BMT patients is from four to eight weeks.

The first successful BMT was done in 1968. Presently, thousands are done each year. The type of procedure chosen (marrow, cord blood, peripheral blood, related or unrelated donor, etc.) depends upon availability and technical factors. There have also been difficulties getting insurers to pay for these treatments. The following Web sites listed have pages with advice for people having trouble getting reimbursement:

- Bone Marrow Foundation: http://www.bonemarrow.org
- International Bone Marrow Transplant Registry: http://www .ibmtr.org/
- National Marrow Donor Program: http://www.marrow.org/
- Blood and Marrow Transplant Information Network: http:// www.bmtinfonet.org/

56.

Ancient and Early Surgery

Columbia University's medical school is called the College of Physicians and Surgeons, reflecting a past in which these healing traditions—those of physicians and surgeons—developed independently, if not, upon occasion, antagonistically. Even today, in England some surgeons proudly call themselves *Mister*, rather than *Doctor*, which they have all earned.

The first surgery performed on the human body probably occurred shortly after the invention of sharp instruments. Evidence of early surgery dates the practice back well into the Neolithic Age (about 10,000 to 6000 BCE).

Even primitive brain surgery was successfully being performed as early as 8000 BCE. Trepanation, during which a small hole is drilled in the skull to relieve pressure on the brain, was a fairly common practice at that time. A piece of the skull (frontal, parietal, or occipital bones) was removed from a living patient to expose the dura mater—the tough fibrous membrane forming the outer envelope of the brain. Even without anesthesia or antisepsis, advances that would be thousands of years away, a patient had a good chance of surviving without brain infection if the dura matter was not breached.

By 2500 BCE the Egyptians, as documented in many carvings, were successfully performing surgical circumcision on both men and women. Using willow leaves and bark to stave off infection, the Egyptians even performed lithotomy, the removal of stones from the bladder, as well as various amputations. Egyptian medical texts offer detailed instructions for many surgical procedures, including the repair of broken bones and suturing of substantial wounds.

The Hindus of ancient India also used surgical procedures to treat broken bones, as well as to remove bladder stones, tumors, and even infected tonsils. But they are most often credited with advances in plastic surgery as early as 2000 BCE, as treatment for the common punishment of cutting off a person's nose or ears for criminal offenses.

The oldest Indian treatise dealing with surgery is the *Shushruta-Samahita*. Concentrating mostly on rhinoplasty ("plastic" nasal surgery) and, amazingly, on ophthalmic procedures like cataract removal, the *Shushruta* described surgery under seven categories: excision, scarification, puncturing, exploration, extraction, evacuation, and suturing. Below is a translated passage from the guide:

> When a man's nose has been cut off or destroyed, the physician takes the leaf of a plant which is the size of the destroyed parts. He places it on the patient's cheek and cuts out of this cheek a piece of skin of the same size (but in such a manner that the skin at one end remains attached to the cheek). Then he freshens with his scalpel the edges of the stump of the nose and wraps the piece of skin from the cheek carefully all around it, and sews it at the edges. Then he places two thin pipes in the nose where the nostrils should go, to facilitate breathing and to prevent the sewn skin from collapsing. There after he strews powder of sapan wood, licorice-root and barberry on it and covers with cotton. As soon as the skin has grown together with the nose, he cuts through the connection with the cheek.

The Greek physician Hippocrates (see advance 3) continued the development of surgical procedures, publishing detailed descriptions

Fig. 39. Descriptions of ancient and early surgery emphasizing the painful and unsterile conditions as contrasted with early modern surgery (*lower left*).

of various surgical procedures in the fourth century BCE. But it was not a task for the weak or timid. Surgical instruments were made of iron, copper, or copper alloys. These blunt instruments were used to remove stones from the bladder as described in the Hippocratic books (which is ironic, considering that the Hippocratic Oath prohibits physicians from performing surgery: *"I will not cut, even for the stone, but will leave such procedures to the practitioners of that craft"*). The ancient Greeks inserted a hollow metal tube through the urethra to empty the bladder. The tube came to be known as a catheter; it was made of copper or lead, straight for women and S-shaped for men. The procedure, like all surgery of the time, was both painful and dangerous.

"A surgeon should be youthful," the philosopher Celsus wrote. "With vision sharp and clear, and spirit undaunted; filled with pity, so that he wishes to cure his patient, yet is not moved by his cries to go too fast or cut less than is necessary, but he does everything just as if the cries of pain cause him no emotion."

By the Middle Ages (fifth century to fourteenth century CE), surgery had been relegated to the "labor" class. These years marked an extended suspension of the practice by trained doctors, as surgery was seen as inferior to medicine. Surgery was performed only by barbers. Incredibly, the same men who traveled the land cutting hair were now trusted to remove tumors, pull teeth, stitch wounds, and drain patients' bodies of substantial quantities of blood in an effort to cure various illnesses. Interestingly, the red-and-white striped pole still identified with barbershops today originally symbolized the red blood and white bandages of this early practice.

It wasn't until the French surgeon Guy de Chauliac published *Chirurgia Magna*, translated as "Great Surgery," in 1316 that surgery was invited back into the fold of respected medicine; but not everywhere.

The next three hundred years were responsible for revolutionary discoveries that advanced both medicine, in general, and surgery, in particular. The French surgeon Ambroise Paré developed a technique called ligating, in which arteries are tied off to control bleeding. This ended the more primitive practice of cauterization, where the bleeding arteries are burned with a hot iron or boiling oil. A string of discoveries, involving the understanding of blood circulation and the existence of tiny blood vessels called capillaries, further increased surgery's effectiveness and efficiency.

But up until the 1840s, surgeons rarely delved deep into the human body or tampered with vital organs because of the intense pain experienced by the patient and the mortal risk of infection. Then, on March 30, 1842, Crawford Williamson Long, MD, removed one of two tumors from the neck of Mr. James Venable using ether as anesthesia (see advance 35). This is the first surgical use of anesthesia, although Dr. Long did not publish his achievement until 1848, resulting in William Morton's receiving credit for the breakthrough.

With the pain of surgery minimized, the only thing holding it back from truly revolutionizing the treatment of countless illnesses was the eradication, or at least reduction, of the risk of infection. Enter Louis Pasteur (advance 15). His discovery that fermentation or putrefaction—the decay and death of body tissue—is caused by bacteria in the air was the first step. British surgeon Joseph Lister's (advance 46) application of Pasteur's work to surgery in 1865 was the final element for bringing surgery into the modern age. Lister, whose development of antiseptic included using a carbolic acid spray to kill germs in the operating room before surgery, is the namesake of the mouthwash Listerine. Later, doctors such as the Austrian Ignaz Semmelweis (advance 31) and American Oliver Wendell Holmes took medical germaphobia one step further, insisting that physicians wash their hands and change into clean clothing immediately before surgery to prevent the introduction of harmful bacteria.

The Development of
Medical Subspecialties

In the twentieth century the expansion of knowledge in all fields
of medicine brought about the development of subspecialties such
as cardiology, gastroenterology, nephrology, hematology, pulmonary medicine, rheumatology, dermatology, endocrinology,
and others. Along with this trend came subspecialty societies and
journals, and more focused research. This has stimulated great advancement, but it has created a number of problems.

As subspecialties define themselves, and separate themselves into
divisions within departments of medicine, and then into autonomous
departments, the relative isolation limits understanding of the integration of the human organism. The best example is in the field of
neurology. Most divisions of neurology have become departments of
neurology, and this has brought about focus and progress. Within
these departments, neurologists can focus on understanding and
treating strokes, or seizure disorders, or cognitive problems, or other
areas of neurologic disfunction. But the price has been that increasingly more internists know very little neurology, and neurologists

know little internal medicine. The same development, spurred on by financial concerns, is now occurring in cardiology. Moreover, many institutions now have departments of internal medicine *and* departments of general medicine (for nonspecialized general practitioners).

Subspecialization has been a necessary phenomenon, brought on by real concerns about the rapid expansion of information, which was later encouraged by economics. The big-ticket procedure subspecialties (cardiology and gastroenterology) tend to want to manage their own incomes, rather than support the department of medicine. This presents real problems in teaching and learning and administering institutionalized patient care. It is one of the reasons why being a chair of medicine is an increasingly difficult job.

But this can change so that the advantages of subspecialization even further outweigh the disadvantages. In all areas, medicine is rapidly incorporating the tools of the information revolution. In the time it used to take me to get to the elevator at my office, I can now have before me most of the world's literature on any subject. And, of course, I can do that at the bedside with young physicians, patients, nurses, and medical students. This is changing the game. We can all learn faster, remember more, and better integrate our knowledge. The rapid availability of information enables clinical physicians to be broadly competent in most areas of internal medicine. The pressure to subspecialize should thus become more purely economic. And everyone but the insurance industry and HMOs agree that the economic organization of healthcare must change.

The Development of
Surgical Subspecialties

S urgical subspecialities have developed along lines similar to medicine (see advance 57), with similar results. Neurosurgery, cardiothorasic, orthopedic, trauma, general (abdominal), urologic, and other branches of surgery have all individually advanced. From our perspective, subspecialization has been a more positive element in the advancement of surgery. This is because the intellectual requirements of practice must be matched with physical skills that take time and effort to acquire.

The information revolution is having the same impact on surgery as in medicine, but I believe its transformative powers will be less profound in surgery because the underpinnings of subspecilization are more vital. This is linked to the fact that surgical training takes so much more time.

59.

Minimally Invasive Surgery

Until the final decades of the twentieth century, almost every surgical procedure required a relatively large incision to expose an operative field in which work could be done. This often involved cutting through the skin and subcutaneous tissues, through underlying muscle layers, and through linings covering the organ cavities— such as the peritoneum in the abdomen, the pleura in the thorax, and the bursas of the joint spaces. Then, the organs to be operated upon might have to be entered, for example, by cutting through the wall of the colon to remove a polyp, or through a bronchus to biopsy a lung tumor. While this is still the case for many operative procedures, operations increasingly are being performed through advanced endoscopic equipment requiring no incision or only tiny incisions to allow entrance of endoscopes. An endoscope is an instrument through which one can gain access to an organ in order to observe and manipulate its contents. The most familiar endoscopes are the otoscope and the ophthalmoscope with which the doctor examines your ears and eyes. Both of these instruments have lenses and light sources, features that are common to almost all endoscopes. With these methods there

239

are no big steel retractors pulling apart the edges of the surgical wound, and the surgeon's hands are outside the patient's body; sometimes they're on another continent, when these instruments are manipulated remotely.

The modern era of endoscopic surgery was ushered in by the development of flexible fiber-optic instruments to examine the upper and lower gastrointestinal (GI) tract. Early attempts at endoscopic examination of the stomach (gastroscopy) were severely limited by the long rigid tube that had to be swallowed and the dim lighting that could be achieved. This required an act of courage on the part of the patient, and it resulted in a limited examination. It was rarely done. Similar difficulties restricted the use of rigid colonoscopes to the last twenty-five centimeters of the colon.

The application of fiber optics to endoscope design led first to a revolution in clinical gastroenterology, then to sweeping changes in pulmonary medicine, general surgery, and orthopedic surgery. By the late 1960s longer flexible instruments with powerful light sources were being used to examine the upper GI tract from the mouth through the duodenum; thus the entire colon could be negotiated and seen clearly. With fiber optics, the long snakelike endoscope could be painlessly inserted, passed to nearly any distance, and bent in any direction. Moreover, light could travel up the fibers producing an undistorted image at the examiner's eyepiece. Through channels in the walls of these instruments, one can apply suction or pass slim cables with small tools attached to the ends, including forceps, brushes, cautery wires, stents, balloon dilators, and lasers—virtually everything required to do surgery.

By the 1980s fiber optics were replaced by video imaging. This allows larger, clearer images that are projected on video screens and can easily be recorded for consultation and teaching purposes. We now have thinner, more flexible, more powerful endoscopes of various kinds through which many operations can be done in the abdomen (laparoscope), the pelvis (culdoscope), the chest (thoraxoscope), or the joints (arthroscope).

Among the advantages of minimally invasive surgery are more rapid, less painful recovery and reduced scaring and complications associated with wound healing.

60.

Birthing Care

To give birth is a fearsome thing . . .

—Sophocles (497–406/5 BCE), *Electra*, 1:770.

Maternal and infant mortality may not tell us everything, but they are a worldwide barometer of health and healthcare. Birth is the essential condition. Everyone experiences it, and with all of its potential for excitement, joy, and pain, it has always been a hazardous time when life and death may most unkindly meet. The way individuals and societies deal with this achingly vulnerable experience provides a no-nonsense perspective on gender relations as well as the political economy, among many vital concerns. So it should not surprise anyone that there have long been struggles over the conduct and control of birthing care.

Fig. 40. Portrait of Louise Bourgeois, who wrote the first midwife manual.

Fig. 41. Pregnant woman with depictions of childbirth from (*top left-clockwise*) Aztec, Japanese, Assyrian, and medieval European cultures.

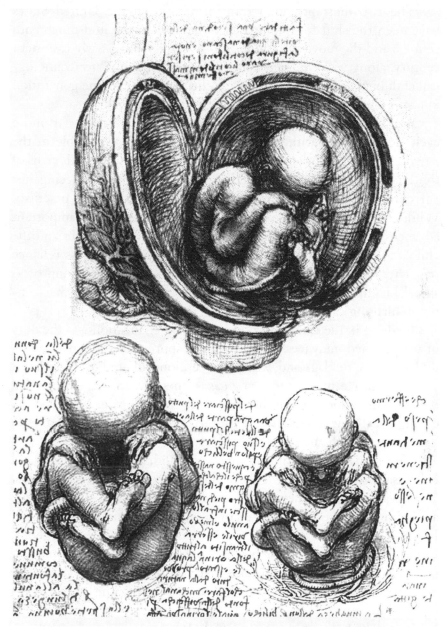

Fig. 42. Drawing of a fetus by Leonardo da Vinci.

The two most profound transformations in the United States have been the shift from midwives to doctors in the mid-nineteenth century and the move from homes to hospitals in the early twentieth century, along with the increasing shift from "natural" to technological childbirths. This all took place in little more than one generation in a stumbling march of progress that is far from over.

But in many ways the broader evolution has been far more cyclical. Beginning with the most "natural" process possible in the days before modern science to the largely chemically assisted births of the late nineteenth and twentieth centuries, we may be seeing our current trends as a movement back toward more traditional practices. While there is no doubt that there have been many important advances in childbirth that serve to protect both the mother and the child, it seems noteworthy that midwifery, home birth, less reliance on general anesthesia, and the introduction of doulas (nonprofessional birthing aids) are all part of an emerging, more mother-centered, birthing care.

Birthing in the earliest periods of human civilization was the duty of well-trained midwives. These women apprenticed for many years under intense tutelage and, without exception, adhered to the strictest standards of cleanliness. In many cases, their practice even included the administration of primitive pre- and postnatal care. Male doctors during this time considered childbirth to be beneath them and assisted in only the most extreme cases.

Unfortunately for both mother and child, the Middle Ages and Renaissance periods brought about the near extinction of the midwife (see advance 91). Women were forbidden from practicing medicine, and midwives were often accused of being witches and were executed. In this process, their practices were transferred to barber-surgeons, who came to monopolize the business of childbirth. This transfer, history would reveal, was a mistake that sacrificed the lives of untold women and babies. Many women today express the desire for a woman in attendance, someone experienced in the practice of birthing, which does not mean to exclude men.

Hospital stays during the Renaissance and into the later nineteenth century were almost a death sentence for expecting mothers.

Both doctors and medical students—equally lacking in the experience that had rendered midwives so effective—would regularly march straight from the dissection tables to the labor wards to examine expecting mothers without washing their hands, transmitting diseases not only from the dead bodies to the initial patient but also from patient to patient as the doctor went from internal examinations of one laboring woman to another. More women died from childbed fever than survived during this period of ignorance, and a large percentage of newborns died within forty-eight hours of birth.

The situation became so dire that many hospitals began to employ midwives—those that had not been accused of witchery—to deliver babies of the lower classes in a separate wing. As a result, the incidents of childbed fever among the lower classes all but disappeared. Before long, affluent woman began to demand the services of midwives in order to avoid death.

Incredibly, it wasn't until the end of the nineteenth century that the medical community was dragged kicking and screaming, by Dr. Ignaz Semmelweis, to the connection between a lack of cleanliness and the spread of infection (see advance 31).

The next major change in birthing care was the introduction of anesthesia during labor. This happened in the 1840s, but concerns about its effects on mother and child health, and on their relationship, made this practice controversial. Inhalation of the gas substantially reduced the pain experienced by the mother, but she would not be awake to greet her newborn. The clergy was also against it, citing references to the pain of childbirth in the Bible. But on May 14, 1853, Queen Victoria gave birth to Prince Leopold with the assistance of this controversial anesthetic, and the world took notice. The growing popularity of this practice is what precipitated nursery wards in hospitals, because the new mothers were rendered unable to care for their babies until the effects of the gas wore off.

However, at least in the United States, childbirth was still an alarmingly dangerous endeavor. At the turn of the twentieth century, approximately 124 out of every 1,000 babies died during childbirth in this country, prompting the Children's Bureau in Washington, DC, and the Maternity Center Association in New York City to look to foreign countries with significantly lower infant mortality rates for guidance. What they found was that the United States was one of the few countries without nurse-midwife participation in the process of childbirth. In 1929, more than two thousand years after the Bible documented the practice, a nurse named Mary Breckinridge brought North America its first professional midwife from Britain. By 1953, when Dr. Fernand Lamaze published his revolutionary system of controlled breathing and self-hypnosis, there were seven accredited educational programs for midwifery in this country. Today there are more training programs, and there are over seven thousand certified nurse-midwives practicing in the United States, assisting in nearly 10 percent of all US births. Yet despite advanced technology and the fact that 99 percent of American babies are born in hospitals, the United States still ranks twenty-sixth worldwide in infant mortality.[87]

At the beginning of the twentieth century, for every 1,000 live births, 6 to 9 women in the United States died of pregnancy-related complications, and approximately 100 infants died before age one. From 1915 through 1997, the infant mortality rate declined more than 90 percent to 7.2 per 1,000 live births, and from 1900 through 1997, the maternal mortality rate declined almost 99 percent to less than 0.1 reported deaths per 1,000 live births (7.7 deaths per 100,000 live births in 1997).[88] Environmental interventions, improvements in nutrition, advances in clinical medicine, improvements in access to healthcare, improvements in surveillance and monitoring of disease, increases in education levels, and improvements in standards of living contributed to this remarkable decline. Despite these improvements in maternal and infant mortality rates, significant disparities by race and ethnicity persist, and are growing.

Among the greatest advances in birthing care has been the realization that the quality of life of parents is a primary determinant of the success or failure of the birthing process. We now recognize that birthing care must address prenatal, natal, and postnatal issues. This notion, while generally accepted, has clear implications with respect to social welfare and the conditions facing everyone. About a decade ago, a paper was published which documented that a boy born in Harlem had a shorter life expectancy than a boy born in Bangladesh.

Similarly, black, American Indian, and other babies of the disenfranchised now have about a 2.6 times greater chance of dying in their first year of life than white babies. The gap in maternal mortality between black and white women has increased since the early 1900s. During the first decades of the twentieth century, black women were twice as likely to die of pregnancy-related complications as white women. Today, black women are more than three times as likely to die as white women.[89]

Well-designed government programs can work wonders, and Medicaid produced marked reductions in maternal and infant mortality. The improvements in birthing care have been glorious, a real triumph, but we still face challenges. The lens of medicine exposes the problems, but the actions of people must continue to provide the solutions.

61.

Hospitals

Health care is being converted from a social service to an economic commodity, sold in the marketplace and distributed on the basis of who can afford to pay for it.

—Dr. Arnold Relman, editor of the *New England Journal of Medicine*, criticizing the takeover of public hospitals by commercial businesses

Fig. 43. Hôtel-Dieu, the first hospital in Paris.

Many people believe that Jesus Christ and the altruism of Christianity are solely responsible for the creation of institutional healing centers. Hospitals, this theory holds, did not exist before the year 1 CE. And while this view is historically inaccurate, it is also interesting and informative.

There certainly is an irrefutable link between religion and the development of institutionalized healing. But religion and altruism did not begin with Jesus. In other words, this creation myth, at least in relation to the invention of hospitals, is not wrong so much as it is simply off by about forty centuries.

As early as 4000 BCE, religious groups had constructed both physical and organizational structures for the purpose of healing the physically sick. The temples of Saturn in Egypt, and later of Asclepius in Asia Minor, functioned as hospitals in addition to what we now consider medical schools. It is even believed that the temple of Saturn included an asylum for the mentally insane. But these facts alone are not enough to simply write off claims by some to Christianity's invention of the hospital. Revisionist historians can successfully argue that these pagan institutions were more gospel than hospital, as much of the work was done by "healer-gods," who would simply admit the sick and wait for healing instructions from God in a dream. So where and when were the first institutions that would conform to the modern definition of a hospital created?

As in much of early medicine, one must look to the Far East for answers. Bramanic hospitals, in which modern-leaning, scientific approaches were implemented, were established in Sri Lanka as early as 431 BCE. And by 230 BCE, King Ashoka had founded eighteen hospitals in Hindustan that were subsidized by royal funds and staffed with physicians as well as nurses. These, by any definition, were the precursors to the modern hospital.

And it took another three hundred years before the concept migrated west. By 100 BCE, the Romans had established the concept of the *valetudinarian*. Constructed as a place to care for sick and injured Roman soldiers, a valetudinarian was a huge advance in modern medicine. But even at this late date, Romans considered the diseased damned by the gods, and as a result the role of these early hospitals was as much isolation as care.

The Western world's first modern concept of a hospital dates from 331 CE when Constantine made caring for the sick a responsibility of the Christian church. In 370 St. Basil of Caesaria established a religious foundation in Cappadocia that included a hospital, an isolation unit for those suffering from leprosy, and buildings to house the poor, the elderly, and the sick. Following this example, similar hospitals were later built in the eastern part of the Roman Empire. St. Benedict at Monte Cassino, founded early in the sixth century, became famous as a place where caring for the sick was placed above every other Christian duty.

When the Hôtel-Dieu of Lyon was opened in 542 and the Hôtel-Dieu of Paris in 660, more attention was given to the well-being of the patient's soul than to curing bodily ailments. The monasteries of France often included an *infirmitorium*, a place to which their sick were taken for treatment. These institutions frequently included a pharmacy and even a garden with medicinal plants. In addition to caring for sick monks, the monasteries opened their doors to pilgrims and other travelers.

Religion continued to be the dominant influence in the establishment of hospitals during the Middle Ages. The growth of hospitals accelerated during the Crusades, which began at the end of the eleventh century. Military hospitals came into being along the traveled routes. The Knights Hospitalers of the Order of St. John could care for as many as two thousand patients, and it became famous for its treatment of eye disease.

The twelfth century saw hospitals established in Baghdad and Damascus. Arab hospitals admitted patients regardless of their religion, race, or social order. In the Middle Ages, which were dominated by religion, the Benedictines created more than two thousand monastic hospitals. But, just as important, the Middle Ages saw the beginnings of secular institutions' influence on the creation of hospitals. Many cities and towns of Europe supported institutional healthcare by the end of the fifteenth century. Then, when Henry VIII dissolved the monasteries in 1540, secular authorities took the primary responsibility for the care of the sick, injured, and handicapped. The first such hospital in England was established in 1718 by Huguenots from France. That was almost immediately followed by the founding of London's Westminster Hospital in 1719, Guy's Hospital in 1724, and the London Hospital in 1740. Between 1736 and 1787 hospitals were established outside London in at least eighteen cities. The initiative spread to Scotland with the opening of Edinburgh's Little Hospital in 1729.

North America's first public hospital, still in existence today, was built in Mexico City in 1524 by Hernán Cortés. The French established the Hôtel-Dieu du Précieux Sang in 1639 in Quebec, which is still in operation at a different location in Canada. Four years later, a French noblewoman named Jeanne Mance built a hospital on the island of Montreal. This institution gave birth to the Hôtel-Dieu de St. Joseph, out of which grew the order of the Sisters of St. Joseph, now considered the oldest nursing group organized in North America.

It was almost a quarter century later before the first hospital in the territory of the present-day United States was born. That happened in 1663 on Manhattan Island. These early hospitals were primarily almshouses, constructed to give the needy and the elderly a safe place to maintain their independence. One of the first was established by William Penn in Philadelphia in 1713. And the first incorporated hospital in America was Philadelphia's Pennsylvania Hospital, which obtained a charter from the crown in 1751.

Hospitals represent a sanctuary for the sick. They have been the site of many advances because the gathering place of healers and sick people creates a laboratory for observation and innovation. Think of the Fool of Pest (see advance 31). But the example of Semmelweis also instructs us that hospitals can be hazardous places as well. Until recent times they remained essentially unchanged, with large wards containing many beds. In the 1960s, when I was a medical student and then a house officer, most of the in-hospital patients were housed in large rooms containing eight or sixteen beds lined up around the periphery. There was no privacy at all, and monitoring was done by direct observation; radios and later televisions were considered very disruptive; all the trials and tribulations that are inherent in the hospital experience were lived out in public. Since then, the revolution in electronics and miniaturization has allowed vast improvements in patient privacy and comfort. This has influenced the basic structure of the hospital.

But hospitals are very complex institutions and change comes slowly. They still tend to be large, forbidding, mazelike, impersonal, and understaffed places, where the pace is frenetic and efficiency is compromised. Nevertheless, in my experience, dedicated care and excellent outcomes are commonplace. The challenge is to create an environment that is more welcoming and engaging, while increasing the precision of patient care. For the past decade hospitals have functioned in an atmosphere of anxiety, with sudden closings, forced mergers, and other major structural changes driven by marketplace concerns. This has not moved hospitals in the patient-oriented direction that would be more beneficial to the vast majority of patients and healthcare professionals. The ever-increasing salaries of hospital administrators in the face of hardship within the hospital reflects the effect of corporatization of the American hospital.

The Department of Veterans Affairs is trying to "streamline" by eliminating many of its hospitals and possibly selling them off to the private sector. And every day in big cities and small towns you can hear stories like this one from Plattsburg, New York:

> Hospitals throughout the North Country would lose millions under health-care cuts and new taxes proposed by Governor George Pataki, according to a study by the Healthcare Association of New York State. The losses would result from Pataki's proposals to add a 0.7-percent tax on hospital revenues and cut reimbursements for health services for the poor and the elderly.[90]

The great advance that hospitals represent, and their necessity, should be clear. They should be improving, not struggling to stay alive in a tenuous marketplace. Hospitals should conform to Robert Frost's description of home: "the place where when you have to go there, they have to take you in." I am afraid that for some people there will not be any "there" there.

Coronary Care and
Other Intensive Care Units

Specialized units where critically ill patients receive intensive care save thousands of lives each year. The first modern intensive care unit was established in 1958 at the Baltimore City Hospital by Dr. Peter Safer. More than five million people are admitted to the intensive care units (ICU) every year in America. The fact that 10 percent of them die there is evidence of how sick they are; about 90 percent leave the units improved.[91]

Patients receiving medical care in intensive care units account for nearly 30 percent of acute care hospital costs, yet these patients occupy only 10 percent of inpatient beds.[92]

The ICU is generally the most expensive and resource-intensive area of medical care. In the United States estimates of the 2000 expenditure for critical care medicine ranged from $15 billion to $55 billion. This is about 0.5 percent of the GDP and about 13 percent of the national healthcare expenditure.[93]

The most important resource of intensive care units are the specially trained and dedicated nurses and doctors who master the constantly upgraded equipment and special techniques that are introduced on a constant basis.

One of the pioneers of the ICU, particularly the coronary care unit (CCU), was Bernard Lown. He invented the defibrillator in the 1960s, introduced the heartbeat-regulating drug Lidocaine, and pioneered studies on the role of psychological and behavioral factors in heart disease. Lown also won the 1985 Nobel Peace Prize with a Soviet physician for their work on nuclear nonproliferation. He created SatelLife in 1987, which uses satellite and Internet technologies to bring health information to developing countries. SatelLife oversees ProCOR, which provides health workers in developing countries information on combating an emerging epidemic of heart disease.

63.

Streptomycin and Control of Tuberculosis

Evidence uncovered from the fossil record shows that our earliest ancestors suffered from tuberculosis. Once, TB was feared the way AIDS is now. Doctors who developed the disease immediately lost their livelihood. Control of tuberculosis came about in the mid-twentieth century, but by the end of the century, antibiotic-resistant tuberculosis had become a very serious problem (see advance 65).

Selman Abraham Waksman was born in 1888 near Kiev, Russia. In 1911 he entered Rutgers University on a state scholarship, and by 1916 he had earned a bachelor's and a master's degree in agriculture. In 1918 he became a naturalized US citizen and was appointed a research fellow at the University of California where he received a PhD in biochemistry. Over the next two decades, he became one of the world's foremost authorities on soil microbiology. After the discovery of penicillin, he was a major player in initiating a calculated, systematic search for antibiotics among microbes. In 1943 he isolated the antibiotic streptomycin, which became the first agent in the treatment of tuberculosis. This discovery was rewarded with the 1952 Nobel Prize for Physiology or Medicine. He is also credited with coining the term *antibiotic*, meaning "against life," for chemicals that destroyed microorganisms harmful to humans.

Streptomycin is derived from a soil bacterium of the genus *Streptomyces*, and it is active against both gram-positive and gram-negative bacteria, including species resistant to other antibiotics. Streptomycin is still a mainstay of tuberculosis therapy. Because resistant tubercle bacilli emerge during treatment, antibiotics are usually used in combination with one or more of the drugs isoniazid (INH), ethambutol, or aminosalicylic acid. In the 1950s, INH brought about the closure of TB hospitals because, in addition to being highly effective, before the organisms are killed, just a coating with the drug makes them noninfectious. Multiple-drug-resistant TB has brought about observed treatment programs to be sure patients take their medications and are less of a danger to others. Streptomycin acts by inhibiting protein synthesis and damaging cell membranes within susceptible microorganisms.

Millions of lives have been saved by the brilliant research efforts of Selman Waksman and his colleagues.

64.

Digitalis

The heart fails when its pumping action is impaired. The consequences are frequently immediately apparent on physical examination, and they are felt by the patient as symptoms. These symptoms result from events taking place both upstream and downstream from the heart (see advance 10). Forward failure refers to inadequate delivery of blood to organs like the brain and the kidneys (left ventricular failure). Backward failure refers to the accumulation of blood downstream, or behind the heart, accumulating in the lungs, the legs, and the abdomen as edema fluid, or what was formerly known as dropsy (right ventricular failure). Among the symptoms of congestive heart failure (CHF) are fatigue, confusion, shortness of breath, and swelling of the legs and abdomen. Early or mild CHF may have minimal symptoms and signs, but it generally progresses to become obvious and debilitating.

Many processes that cause the heart to increase its workload for a prolonged period or produce anatomic damage that makes it more difficult for the heart to function (damaged valves) result in CHF, which is one of the most common causes of death. These processes

Fig. 44. The leaves of the beautiful foxglove give us digitalis for stimulating the heart.

include high blood pressure, toxic injury, infections of the heart, and congenital or genetic abnormalities. So CHF has been a major problem for a long time, and as the population ages, it increases.

The two main approaches to treating CHF are to reduce the work of the heart by lowering elevated blood pressure, fixing faulty valves, or removing retained salt and water (diuresis), and to increase the force of contraction of the heart muscle. It was the leaf of the beautiful foxglove plant (containing digitalis) that improved the function of the heart. Foxglove became a mainstay of treatment by the late eighteenth century, was a miracle drug when there were few effective treatments, and is still in use.

In 1785 William Withering published *An Account of the Foxglove and Some of Its Medical Uses etc.; With Practical Remarks on Dropsy and Other Diseases.* There he described digitalis's powerful stimulatory action on the heart and how it can reduce the edema of heart failure. He had heard of the treatment from a Shropshire woman who made an herbal tea using "a secret family recipe" to relieve swollen legs. The potion contained over twenty ingredients, but Withering surmised that foxglove leaf (digitalis) was the active ingredient. On December 8, 1775, he began treating a fifty-year-old man with shortness of breath and dropsy using the foxglove tea. The patient began to make more urine, his breathing became easy, his abdominal swelling subsided, and in about ten days he was eating with good appetite and feeling well.

Digitalis is used to treat both CHF and rapid cardiac rhythm (tachycardia). While the dry digitalis leaf preparations are still available, their dosimetry is tricky, so they are dangerous. The most common preparation used today is digoxin, which has significant potency for the treatment of heart failure as well as atrial fibrillation and other supraventricular tachycardias (rapid heart rates arising above the ventricles in the atria). Careful attention to dosage and, if needed, drug level monitoring (determined by RIA, see advance 41) is used to help ensure that digitalis toxicity (a common side effect) does not occur. Toxicity may include visual difficulties as well as cardiac arrhythmias (it both improves arrhythmias and, at higher blood levels, can cause them). These may be at times life threatening. Fortunately, a specific antidote (Digibindâ) is available for treatment of serious digitalis toxicity.

Digitalis's mechanisms of action are complicated and poorly understood. But the improvement that it brings to millions of people makes it one of our greatest advances.

65.

The Pneumococcus from Hell and Understanding Antibiotic Resistance

Human industry is having a profound effect on the ecosystems of this planet. Though it's bringing technological advances, it is also degrading the world in which we live. The struggle to keep the good while rejecting the bad requires information, intelligence, and a distribution of power that does not concede everything to corporate profit. Most of us are most disturbed by things that we can see, smell, and feel. The stench of fumes, the accumulation of garbage, the pollution of rivers, the clog of traffic—these are tangible, inescapable. But on all the surfaces of the earth, in the waters, underground, and inside our bodies, the changes we have wrought over the past sixty years in the invisible microbial world have distorted natural evolution to the extent that we may be facing disaster. Not in the future—now.

Since the introduction of antibiotics over sixty years ago, we have enjoyed increasing mastery over microbial disease (see advance 34).

Fearsome killers like smallpox (see advances 16 and 23), rheumatic fever (see advance 34), and tuberculosis (advance 63) were either eliminated or reduced to routine problems, hardly requiring hospitalization, with very high rates of cure. In recent years we have developed ever-more-powerful antibiotics with activity against an increasingly broad spectrum of microbes, and we have provided them to billions of people all over the world. This practice has been considered both good medicine and good business. But it has resulted in a startling phenomenon: the adaptive development, through natural selection, of genes for antibiotic resistance by virtually every bacterial pathogen. In other words, the microbes are fighting back.

The Rockefeller University Workshop on Multiple-Antibiotic-Resistant Bacteria declared that "[a]fter a half-century of virtually complete control over microbial disease in the developed countries, the 1900s have brought a worldwide resurgence of bacterial and viral diseases."[94] The facts are frightening, and practicing physicians throughout the world are encountering the devastating consequences of the natural selection of bacterial clones that have acquired resistance to multiple antibiotics through stable genetic alterations. In other words, they are immune to a broad selection of antibiotics, so the antibiotics are not working. This selection of resistant clones results from the extensive distribution of broad-spectrum antibiotics for medical purposes and in animal feeds. Clones that have acquired resistance to antibiotics have an obvious selective advantage. Resistance may be acquired through a mutation in a microbe's genome which codes for an enzyme that attacks the antibiotic's chemical structure. And the resistant clones have spread around the world like wildfire because of the crowding of populations in ghettos and shelters, the movement of people through travel and immigration, the growing number of immunocompromised individuals, and the increasingly global nature of the food supply.

Although resistance was first observed shortly after the introduction of antibiotics in the 1940s, it was, until recently, a minor problem involving a limited number of organisms for which alternative antibiotics were effective. Currently, resistant staphylococci make up more than 90 percent of isolates (organisms isolated from diseased people).

Outbreaks of multidrug-resistant tuberculosis have erupted in almost every state in the United States and throughout the world. In New York City nearly 50 percent of cases in large institutions are multidrug resistant. In contrast to the high cure rate of tuberculosis caused by traditional strains, resistant tuberculosis results in 40 to 60 percent fatality in patients with normal immunity and over 80 percent in immunocompromised patients. Similar problems are developing with virtually all species of pathogens. The earlier "Mission Accomplished" banner flown by the medical community to signal victory over infectious diseases was, at best, shortsighted.

Consider the pneumococcus, one of the most widespread bacterial pathogens. When penicillin was introduced in the mid-1940s, the pneumococcus was uniformly sensitive to low doses of penicillin and other drugs. This allowed a substantial decline in mortality from pneumococcal disease worldwide. But in the last decade multidrug-resistant clones with resistance to penicillin, erythromycin, tetracycline, chloramphenicol, trimethoprimsulfamethoxazole, and third-generation cephalosporins have spread throughout the world. Now, when illness is caused by the "pneumococcus from hell," even a localized middle ear infection may require hospitalization and intravenous treatment with vancomycin, one of the newer antibiotics costing several times the gram-for-gram price of pure gold.

This vital issue, like all others having to do with health, affects everyone and is not beyond the grasp of any informed member of society. Because of overuse of antibiotics, hospitals, where the vulnerable and the infected meet, have become increasingly dangerous places. Consider these facts:

- Nearly two million patients in the United States get an infection in the hospital each year.
- Of those patients, about 90,000 die each year as a result of their infection—up from 13,300 patient deaths in 1992.
- More than 70 percent of the bacteria that cause hospital-acquired infections are resistant to at least one of the drugs most commonly used to treat them.
- Persons infected with drug-resistant organisms are more likely

to have longer hospital stays and require treatment with second or third choice drugs that may be less effective, more toxic, and more expensive.[95]

In short, antimicrobial resistance is driving up healthcare costs, increasing the severity of disease, and increasing the death rates from certain infections.

Dealing with the emergence and global distribution of resistant pathogens will require coordinated action on an international scale. There will have to be new approaches to antibiotic pharmacology. More important, there will have to be changes in medical and industrial practice, including the banning of routine treatment of cattle with antibiotics to increase their weight. This will require *you* to be informed and to demand that antibiotic overuse be stopped throughout society and industry.

66.

The Development of Antiviral Drugs and Cures for Childhood Leukemia

When Gertrude Elion (1918–1999) joined the Wellcome Research Laboratories in 1944, she took a position as assistant to Dr. George Hitchings, with whom she would share the Nobel Prize in 1988. He was a great scientist and a great boss, assigning her to cutting-edge work and encouraging her to expand into whatever intellectual space she could fill. Their partnership produced medical marvels.

George Hitchings gave Gertrude Elion freedom and the respect that would lead to a true scientific partnership. In the 1940s this was almost unheard-of. For this honor she gave up her struggle to get a PhD and threw her immense talent and energy into their work. It was some six years before Rosalyn Yalow (her fellow alumna of then little Hunter College in the Bronx) met Sol Berson (see advance 41). Like Berson and Yalow, the Elion and Hitchings partnership was a triumph of cooperation and productivity. It was not only that Elion was a woman in a man's world, but she only had a master's degree. She was ripe for exploitation or simple underestimation. But this partnership existed on a higher level. To my knowledge, this is all the more unusual because the great work that they did took place within a drug corporation's laboratories. This was not narrow product-driven research—though Elion was inspired to do something about malignant disease because her father had died of cancer.

In 1944 very little was known about nucleic acid structure and function (see advances 26–28), but Hitchings and Elion certainly were aware that they needed to understand more about purines and pyrimidines, the key constituents of DNA and RNA. They knew that nucleic acids controlled cellular events, that heredity was made manifest through nucleic acids, and that viruses were composed of them. And since all cells needed them, and rapidly dividing cells, like bacteria and cancer cells, had to be making lots of nucleic acids, it might be possible to stop the growth of such cells with antagonists to the purine and pyrimidine bases (simple ring structures that make up the nucleic acids). Few of the many enzymes required for nucleic acid synthesis were known at that time, and the pathways for synthesis of purine and pyrimidine bases were not understood.

In 1948 they found that a synthetic purine derivative (2,6 diaminopurine) inhibited the growth of bacteria (*L. casei*) and that the inhibition was reversed by adenine but not by the other natural purines. When tested on the growth of mouse tumor and leukemia cells in tissue culture, diaminopurine proved to be highly inhibitory. When they tried it in human adults with chronic leukemia, it produced two good remissions, but two other patients experienced severe bone marrow suppression (bone marrow has many rapidly proliferating cells). It also showed antiviral activity in animals infected with a DNA virus (vaccinia), but it proved too toxic to try in humans. By 1951 they had made and tested over one hundred purine analogues for antigrowth activity in cultures of *L. casei*, discovering that substitution of sulfur for oxygen at the 6-position of guanine or hypoxanthine produced active compounds. These two synthetic purine analogues, 6-Mercaptopurine (6-MP) and 6-thioguanine (TG), had activity against a wide variety of rodent cancers and leukemias. Then 6-MP was tested in children with acute fatal leukemia. At that time these children were being treated with steroids and methotrexate, and half of them were dead within three to four months; only 30 percent lived for one year. In fact, 6-MP could produce a complete remission in which there were no abnormal cells to be seen in the bloodstream and the children felt healthy, but relapses at various intervals after treatment were common. The median survival time became one year, and a few children stayed in remission for many years. The FDA approved 6-MP in 1953, only two years after it was first synthesized. It was called an antimetabolite of nucleic acid. In other words, it interferes with the metabolism of nucleic acids, and it clearly was the way to go. In short order there were over a dozen drugs with activity against leukemia, and when used in various combinations, with years of maintenance therapy, 80 percent cure rates were achieved by the 1980s. The rates have continued to improve.

This line of investigation focusing on purine analogues, their effects on cells, and then on human illnesses, led this great team to discover other wonder drugs. In the early 1960s they developed azathioprine, closely related to 6-MP, which helped bring about the first successful kidney transplant using unrelated donor organs (see advance 52), and it is still a mainstay of immunosuppressive treatments. Then, they developed allopurinol, which inhibits the primary enzyme (xanthine oxidase) responsible for making uric acid from nucleic acid purines and is very useful in the treatment of the painful and deforming symptoms of gout. And because DNA and RNA are essential for all life-forms, it is easy to see that their approach—that of making and testing chemical analogues that can modify the production of these nucleic acids—produced drugs that worked against bacteria, protozoan parasites, malignant diseases, immune processes, and the first antiviral drugs.

In the 1970s they developed acyclovir (ACV) and found it to be very potent against herpes and related viruses, without harming the mammalian cells that harbor them. In studying ACV's mechanisms of action (inhibition of viral DNA polymerase) the team learned a great deal about viral and host interactions. This has led to the development of other antiviral agents.

And so the marvels of new drugs go on. But there is broad agreement in the medical-scientific community that the current restrictions imposed upon discovery and dissemination of knowledge by corporate ownership of "intellectual properties" is slowing progress. This phenomenon was unknown even in recent times but is now virtually ubiquitous. The marketplace is a fine driving force for technology, for the development of new products. We are not antimarket, but we must transcend this newly imposed narrowness demanded by return on investment. What if they make wonder drugs that most people can't afford? And what if they make scientists into technocrats. The marketplace, in its current unrestrained manifestation, is driving out other theaters of investigation and is a poor driving force for the basic scientific discoveries that make breakthroughs. Even federal research funding is now greatly influenced by the requirement of having narrowly focused, clearly predictable goals. No one uses the term "product-driven" research, but that, in essence, is what it is.

Medical science is now so advanced that it is paradoxical that we face this dilemma. But how do you validate breakthroughs not made, especially in a time of innovations? An often-heard quote in the living rooms where medical-scientists and physicians gather is, "It's the best of times, it's the worst of times."

67.

Chemotherapy for Malignant Disease

> In poison there is physic, and these news,
> Having been well, that would have made me sick,
> Being sick, have in some measure made me well.
>
> —William Shakespeare, *Henry IV, Part 2*, 1.1.137–39

The treatment of cancer and other malignancies with chemical therapy alone, or in combination with surgery and radiation, has brought cures to many sufferers and very significant extention of useful life to many more. And early spectacular successes with methotrexate for the treatment of choriocarcinoma of the uterus and cures for childhood lukemia (see advance 66) made it seem that cancer was going to be conquered in the 1960s. The realization that combinations of tumorcidal chemotherapeutic drugs could improve the results added to the optimism. Indeed, chemotherapy has been a great advance by any measure, but it is not the answer to cancer that many hoped and believed it would be.

Chemotherapy is primarily aimed at the cell's nuclear events related to DNA replication (see advances 27 and 28). Since those events take place in all cells, these drugs are no "magic bullets." They kill cells roughly in proportion to how fast they are replicating. Therein lies whatever specificity they have: tumor cells replicate more

269

rapidly than normal cells. But rapidly replicating normal cells, such as those in bone marrow, the intestinal lining, or hair folicles, will often feel the effects of the cell poisons.

The five main categories of chemotherapeutic drugs—alkylating agents, antimetabolites, plant alkaloids, topoisomerase inhibitors, and antitumor agents—all attempt to affect cell division or DNA synthesis.

Alkylating agents try to stop tumor growth by cross-linking certain DNA double-helix strands, making the strands unable to uncoil and separate and thereby unable to replicate.

Antimetabolites work by pretending to be purine or pyrimidine, which are chemicals that become the building blocks of DNA. Antimetabolites prevent these substances from becoming incorporated into DNA, stopping the normal development and division in an effort to control and kill the cancer cells at a higher rate than the healthy ones.

Plant alkaloids attempt to block cell division by preventing microtubule function. Microtubules are protein structures, tiny tubules, that are found within cells. They are vital for cell division. Blocking them will prohibit the reproduction of cancer cells.

Topoisomerase inhibitors interfere with transcription (see advance 73) and replication of DNA by disrupting proper DNA supercoiling. Without topoisomerases, which are the essential enzymes maintaining the continuity of DNA, the cancer cells cannot reproduce.

Antitumor antibiotics are products of strains of the soil fungus *Streptomyces*. There are two very distinct ways antitumor antibiotics prevent cell division. First, they attempt to bind to DNA by interposing themselves between two adjacent nucleotide bases, making them unable to separate. Second, they inhibit ribonucleic acid (RNA), which in turn prevents enzyme synthesis.

There are increasing numbers of drugs within these classes, and there are new classes of chemotherapeutic agents, many bringing justifiable optimism to patients and oncologists. But this field knows the reason for understatement.

68.

Therapeutic Radiation

Roentgen's discovery of x-rays in 1895 was one of the most dramatic events in the history of medicine (see advance 11). Immediately, x-rays were put to work in the service of medicine as a diagnostic tool. What seems incredible to us is that therapeutic radiation was an almost instantaneous offspring of this discovery. It is strong testimony to the positive interchange between science and commerce.

Within days of its announcement, Roentgen's "new light" was being used for the treatment of both benign and malignant conditions. Radiation therapy was originally defined as the use of high-energy x-ray radiation (later expanded to gamma rays, proton rays, and neutron rays) to kill or shrink localized cancers or other unwanted collections of cells.

The first such application was in Chicago by an electrician and metallurgist named Emil Grubbe, believed to be the first person to use x-ray technology for the treatment of cancer. Grubbe's successful experiment on a fifty-five-year-old patient suffering from recurrent breast cancer was not published for several years and hence did not receive proper recognition. But the proverbial goose was loose. Scientists across the globe would dedicate the next two years (and in many cases, the remainder of their lives) to the development and perfection of this important advance.

Leopold Freund from Vienna is credited with the first use of radiation therapy in benign conditions (the skin lesions of pediatric nevus and lupus vulgaris) in 1898. One year later, in Stockholm, Thor Stenbeck began treating a forty-nine-year-old woman's basal-cell carcinoma of the nose with over one hundred treatments in just nine months; the woman was alive and well thirty years later. In 1901 Boston-based Frands Williams documented the successful cure of cancer of the lower lip using x-rays. And by 1902 the *American X-Ray Journal* listed over one hundred different conditions being treated in the United States.

Over the next thirty years, radiation therapy took its place as both a cure for formerly untreatable diseases as well as a source of symptom relief. The aim is always to kill as many cancer cells as possible, while doing minimal damage to healthy tissues. Early radiation therapy's only limitation in this goal was technology.

Among the limiting factors in early treatment were skin tolerance and the damage to adjacent healthy tissue. In addition, radiation therapy delivered in a single large dose produced severe, life-threatening side effects. In the 1920s the use of ionizing radiation in daily increments for a period of weeks was introduced, and in the 1930s an even safer dosimetry method, known as fractionation, was begun. Fractionation divides the dose into several even smaller increments instead of a single larger daily dose, and it proved so effective that it is still a mainstay of radiation therapy.

Major advances in radiation therapy from 1930 through the1960s took three forms: megavoltage, supervoltage, and conformal radiation. Megavoltage—high-energy x-rays and cobalt-60 gamma rays—concentrated the beam's maximum energy one to two centimeters below the skin surface. By focusing radiation closer to tumor targets, cure rates increased while many complications and harmful side effects dramatically decreased. The increased effectiveness and decreased complications of megavoltage x-rays were further accelerated by the development of supervoltage x-rays, capable of focusing the energy of its beam four centimeters below the skin. Supervoltage x-ray technology became possible only with the discovery of betatron in the 1960s and the development of linear accelerators in the late 1960s and 1970s.

But it was the development of conformal radiation therapy that brought the fight against cancer into the twenty-first century. Utilizing CT scan technology to create a three-dimensional virtual tumor, doctors are able to choose a series of precise angles that will optimally conform to the target while compromising very little of the patient's healthy tissue.

All the radiation therapies described so far have been external-beam radiation. In other words, the radiation is focused from a source outside the body onto the area affected by the cancer. This is the most common strategy—encompassing both photon (x-rays and gamma rays) and particle (electrons, protons, neutrons, alpha particles, and beta particles) radiation—and is administered in a similar way to a common x-ray but over a longer period of time.

Internal radiation therapy, also known as brachytherapy, is another increasingly effective method. The advantage of brachytherapy is that it delivers a high dose of radiation to a small area, even dosages that would be higher than healthy tissues could tolerate if it came from an external beam. Small radioactive wires or pellets, the size of a single grain of rice, are implanted directly into the tumor (interstitial radiation) or into a cavity close to the tumor (intracavity radiation). These wires or pellets can be permanent or temporary. The permanent pellets, known as seeds, can emit radiation for several weeks or months, and because they are so small and cause so little discomfort, they are simply left in place after their radioactive material is used up. The disadvantage of internal radiation, similar to its advantage, is that only a small amount of tissue is treated.

For some forms of cancer, including Hodgkin's disease, non-Hodgkin's lymphoma, prostate cancer, and laryngeal cancer, radiation therapy is the most effective treatment. However, radiation is often used in conjunction with surgery, chemotherapy, or both, because survival rates are greater than those achieved by any single type of therapy used alone. Radiation therapy is especially effective when surgical procedures cannot remove an entire tumor without damaging the function of surrounding organs. In cases such as these, surgeons remove as much of the tumor as safety permits, and the remainder is treated with radiation.

69.

Rita Levi-Montalcini
Discovers Nerve Growth Factor

In spite of, or perhaps because of its most unusual and almost extravagant deeds in living organisms and in-vitro systems, NGF [nerve growth factor] did not at first find enthusiastic reception by the scientific community.

—Rita Levi-Montalcini, 1986

single cell divided until there was you. A fertilized egg cell became two cells, then four, then eight, all the same, with no features even suggesting a nerve cell, a skin cell, a liver cell, much less the complete cellular array that makes up your body. Practical insight into the processes of cellular growth and differentiation, along with some control over them, constitutes one of the greatest medical advances and has provided marvels galore.

From the study of developmental embryology in the 1940s came the discovery of NGF, and then epidermal growth factor (EGF), and then many more growth factors and cytokines, including erythropoietin, which stimulates the growth of red blood cells (erythrocytes) and is advertised prominently on television. The prime mover in this field was Rita Levi-Montalcini, a brilliant and charming woman from a very distinguished family of artists, architects, mathematicians, and intellectuals. She became a physician, although she had an early

Fig. 45. Rita Levi-Montalcini, the discoverer of nerve growth factor.

interest in writing, and her urge to discover could not be suppressed as she herself explains:

In 1936 Mussolini issued the "Manifesto per la Difesa della Razza," signed by ten Italian "scientists." The manifesto was soon followed by the promulgation of laws barring academic and professional careers to non-Aryan Italian citizens. After a short period spent in Brussels as a guest of a neurological institute, I returned to Turin on the verge of the invasion of Belgium by the German army, Spring 1940, to join my family. The two alternatives left then to us were either to emigrate to the United States, or to pursue some activity that needed neither support nor connection with the outside Aryan world where we lived. My family chose this second alternative. I then decided to build a small research unit at home and installed it in my bedroom. My inspiration was a 1934 article by Viktor Hamburger reporting on the effects of limb extirpation in chick embryos. . . .

The heavy bombing of Turin by Anglo-American air forces in 1941 made it imperative to abandon Turin and move to a country cottage where I rebuilt my mini-laboratory and resumed my experiments. In the Fall of 1943, the invasion of Italy by the German army forced us to abandon our now dangerous refuge in Piedmonte and flee to Florence, where we lived underground until the end of the war.

In Florence I was in daily contact with many close, dear friends and courageous partisans of the "Partito di Azione." In August of 1944, the advancing Anglo-American armies forced the German invaders to leave Florence. At the Anglo-American Headquarters, I was hired as a medical doctor and assigned to a camp of war refugees who were brought to Florence by the hundreds from the North where the war was still raging. Epidemics of infectious diseases and of abdominal typhus spread death among the refugees, where I was in charge as nurse and medical doctor, sharing with them their suffering and the daily danger of death.

The war in Italy ended in May 1945. I returned with my family to Turin where I resumed my academic positions at the University. In the Fall of 1947, an invitation from Professor Viktor Hamburger to join him and repeat the experiments which we had performed many years earlier in the chick embryo, was to change the course of my life.[96]

While studying the development of chick embryos, Levi-Montalcini found that if she grafted a mouse tumor (mouse sarcoma) onto the body wall, there was extraordinary growth and development of the chick's nerve tissue. Nerve elements grew rather wildly into the mouse tumor itself, but also from the developing chick nervous system into developing kidneys, thyroid, gonads, spleen, and other tissues. In a series of elegant experiments she and her associates proved that the mouse tumor was elaborating and secreting a small protein (peptide) that acted like a hormone by traveling through the bloodstream to stimulate the growth and development of a target tissue (nerve cells). She called the peptide Nerve Growth Factor. At about the same time, her colleague, Dr. Stanley Cohen, discovered Epidermal Growth Factor. They found these growth-promoting hormones in such disparate and unrelated places as malignant tumors, snake venom, and salivary glands. And as the mouse NGF was active in the chick, so there are other cross-species activities, because biologically active peptide sequences tend to be conserved in evolution.

Injecting neonatal mice with antibody that bound to salivary NGF and blocked its activity prevented the development of the sympathetic nervous system. Nerve cells are most sensitive to the growth-promoting and growth-directing actions of NGF during the earliest stages of differentiation. In the adult mice, while there is still activity, it is much diminished.

Growth factors are peptide hormones (see advance 14) that bind to receptors on the cell surfaces of target tissues, with the primary result of activating cellular proliferation and/or differentiation. Many growth factors are quite versatile, stimulating cellular division in numerous different cell types. Others are specific to a particular cell type.

Cytokines are a unique family of growth factors. Secreted primarily by white blood cells, cytokines stimulate both the humoral (circulating antibodies) and cellular immune responses, as well as the activation of phagocytic cells (white blood cells that engulf and digest pathogens, uric acid crystals, and other harmful agents). Cytokines that are secreted from lymphocytes are termed lymphokines, whereas those secreted by monocytes or macrophages are termed monokines. A large family of cytokines are produced by various cells of the body. Many of the lymphokines are also known as interleukins (ILs), since they are secreted by leukocytes (white blood cells) and are able to affect the cellular responses of leukocytes. Specifically, interleukins are growth factors targeted to cells of hematopoietic origin (arising from bone marrow stem cells). The list of identified interleukins grows continuously with the total number of individual activities now at twenty-two.

A partial list of growth factors and cytokines includes the following:

- Epidermal Growth Factor (EGF)
- Platelet-Derived Growth Factor (PDGF)
- Fibroblast Growth Factors (FGFs)
- Transforming Growth Factors-β (TGFs-β)
- Transforming Growth Factor-α (TGF-α)
- Erythropoietin (Epo)
- Insulin-Like Growth Factor-I (IGF-I)
- Insulin-Like Growth Factor-II (IGF-II)
- Interleukin-1 (IL-1)
- Interleukin-2 (IL-2)
- Interleukin-6 (IL-6)
- Interleukin-8 (IL-8)
- Tumor Necrosis Factor-α (TNF-α)
- Tumor Necrosis Factor-β (TNF-β)

- Interferon-γ (INF-γ)
- Colony Stimulating Factors (CSFs)
- Brain-Derived Neurotrophic Factor (BDNF)

In addition to their current uses for such purposes as stimulating red blood cell proliferation in anemic people following chemotherapy, growth factors and cytokines are being applied to a wide range of issues, including tumor genesis and control.

Suggested Reading

1. Rita Levi-Montalcini, *In Praise of Imperfection: My Life and Work* (New York: Basic Books, 1988).

70.

Hemodialysis

A cute and chronic kidney failure can lead to death if allowed to go untreated for weeks or even just days. Kidney failure causes a condition known as uremia, where toxic substances normally removed by the kidneys accumulate in the body. The symptoms are protean, and virtually every cell in every organ may be affected. Kidney failure is increasing nationally at a rate of 10 to 12 percent each year with more than 230,000 Americans undergoing dialysis annually.[97]

The most common and effective form of dialysis today is hemodialysis. Hemodialysis is the process of filtering uremic substances from the blood of patients by passing it outside of their bodies and through a semipermeable membrane surrounded by a bath; this way substances will equilibrate across the barrier according to their molecular size and charge.

In early Rome and later in the Middle Ages, treatments for uremia (Greek for "urine in the blood") included hot baths, sweat therapies, bloodletting, and enemas. Then, Scottish chemist Thomas Graham, later to be known as the Father of Dialysis, set in motion the process to invent hemodialysis. As a chemist, he was using osmosis and dialysis strictly in chemical laboratories to separate dissolved substances or to remove water from solutions through semipermeable membranes. But he also indicated the potential uses of these procedures in medicine. The medical community did not take notice until German physiologist Adolf Fick published a quantitative description of the diffusion process in 1855. Albert Einstein established the scientific basis for the process fifty years later.

Still, the first historical description using this procedure for medical purposes was published in 1913. John J. Abel "dialyzed" anesthetized animals by directing their blood outside the body and through tubes with semipermeable membranes. Eleven years later, German doctor Georg Haas performed the first dialysis treatments involving humans, but all seven patients died as a result of allergic reactions to the anticoagulant hirudin. Eventually Haas began using heparin instead, which caused substantially fewer complications.

The first successful hemodialysis was performed in 1945 by Willem Kolff of the Netherlands. He used a rotating drum fluid bath to successfully treat a sixty-seven-year-old patient who had been admitted to the hospital with acute kidney failure. She died at the age of seventy-three from an illness unrelated to the kidney failure.

Today, hemodialysis can be done by the patients themselves at home. But fewer and fewer new patients do peritoneal dialysis (PD) or home hemodialysis. In 2002, 92.4 percent of patients went to centers for hemodialysis. A decade earlier, that figure was 83.9 percent; ten years before that it was 82.7 percent. So, even as the population of American patients with kidney failure is increasing, the proportion of those that are caring for themselves at home is falling.[98] Chronic hemodialysis is usually performed two or three times per week. It can be eliminated after a successful kidney transplant (see advance 52).

71.

Restoration of Vision with Cataract and Retinal Surgery

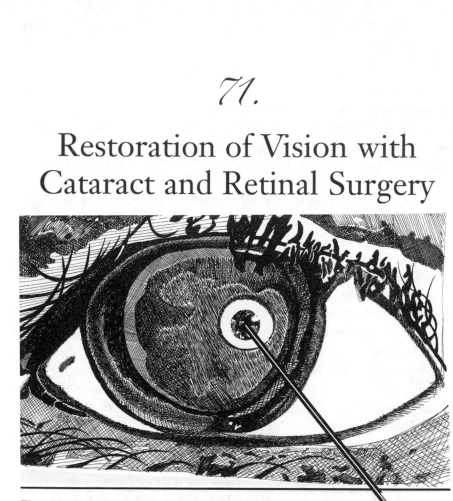

Fig. 46. A drawing of an eye with a cataract (white circle) that has a laser beam focused on it.

Throughout human history blindness has been one of the most debilitating handicaps. Once someone has lost the ability to see, he is immediately and usually permanently reliant on other people in order to participate in society and, in many cases, even just to survive. That reality places advances made in this field on an elevated plane.

Glasses were developed by the Chinese and the Europeans around

1200 CE. Benjamin Franklin, among others, advanced the science greatly, but in practical terms glasses did not become available to the general public until the late nineteenth century. Major accomplishments in the field of vision have been made since, such as contact lenses and intacts, which are permanently implanted in the outer layers of the cornea, but those are both vision enhancers rather than treatments to prevent blindness. For our purposes here, we will focus on the two most common and treatable eye diseases and their surgical treatments: cataracts and diabetic retinopathy.

Cataract surgery is not only the most common form of eye surgery, it is now the most frequently performed surgery in the United States, with over 1.5 million procedures done each year. The entire operation can now be completed in less than ten minutes with the patient returning to normal life immediately. But this was not always the case. For the first several millennia of human existence, cataracts were incurable and the only techniques available to alleviate the symptoms were painful and dangerous. Despite the fact that more than half of all people over sixty will be diagnosed with cataracts, the 1980s marked the first time in human history that a cataract patient's sight could be restored.

A cataract is a clouding of the eye's natural lens, which is located behind the iris and the pupil. This internal lens is made of mostly water and protein arranged in such a way as to keep the lens clear so that light can pass through it. Unfortunately, as humans age, some of the protein may clump together and start to cloud the lens, making it difficult to see. Once a cataract forms, the entire lens must be surgically removed.

In ancient Egypt the surgical technique to treat this blinding condition was called couching. But it should have been called gouging, since the doctor would attempt to push the patient's eye in far enough to dislocate the lens.

Couching remained the procedure of choice for another fifteen hundred years, until Dr. Jacques Daviel surgically removed a patient's cataract in 1748. Unfortunately, there was no anesthetic available at the time and the patient died. It wasn't until the end of the nineteenth century that cocaine began to be used as a topical anesthetic, ushering ophthalmology into a new era. During this time a cataract could be surgically removed, but the patient had to be immobilized for two weeks after surgery with his head wedged between two sand bags. And once the eye was cut into and the inner lens removed, the patient would need extremely thick glasses to see.

Everything changed for the better in 1949 when Dr. Harold Ridley began examining World War II fighter pilots. These pilots often had fragments of plastic from shattered windshields lodged in their eyes but never experienced any adverse effects. From this phenomenon Ridley invented the world's first intraocular lens using the same material that had been lodged in the pilots' eyes. His idea was not only to remove the clouded cataract lens, as had been done before, but also to replace it with a clear, plastic lens of his making. The medical community ignored his proposal, however, and it wasn't until the late 1970s that the use of microscopes in surgery proved Ridley's invention revolutionary. Today, eye surgeons use a flexible, folding lens that fits through a small incision and unfurls once it is within the eye to clear millions of people's vision.

But cataracts are not the only treatable cause of blindness. While they may be the most common eye disease for older people, diabetic retinopathy is the leading cause of acquired blindness among Americans under the age of sixty-five. It is estimated that over fourteen million people in the United States have diabetes mellitus. Approximately half of these individuals have not yet been diagnosed and are unaware that they have the condition.

Diabetic retinopathy is a complication of diabetes in which the blood vessels inside the retina of the eye are damaged. In the early stage of diabetes, blood vessels may become leaky, permitting blood and fluid to seep into the retina. This leakage often swells the delicate retinal tissue, blurring vision. As the blood vessel become increasingly damaged, some of the vessels close, starving the retina of essential oxygen and nutrients.

Laser treatment uses a high-energy light beam to seal leaking blood vessels or remove areas of the retina that have stopped working because their blood supply has closed. This treatment has saved the sight of millions of people who would have otherwise gone blind.

72.

Understanding Retroviruses and the Treatment of AIDS

Science is properly more scrupulous than dogma. Dogma gives a charter to mistake, but the very breath of science is a contest with mistake, and must keep the conscience alive.

—George Eliot (Mary Anne [or Marian] Evans) (1819–1880)

ost genes are expressed as the proteins they encode (see advances 27 and 28). The process by which our genes (DNA) produce proteins occurs in two steps:

Step 1 is *transcription*, in which DNA → RNA
Step 2 is *translation*, in which RNA → protein

Taken together, they make up the "central dogma" of biology:
DNA RNA → protein

The "central dogma" of biology:
5'...A T G G C C T G G A C T T C A...3' Sense strand of DNA
3'...T A C C G G A C C T G A A G T...5' Antisense strand of DNA
→ **Transcription** of antisense strand
5'...A U G G C C U G G A C U U C A...3' mRNA
→ **Translation** of mRNA
Met – Ala – Trp – Thr – Ser – Protein

This is the workaday magic of living cells, the "shéer plód [that] makes plough down sillion shine."[99] It is the common flow of events in living things.

The genome of some viruses that infect animal cells consists of RNA, yet they may enter our cells, join our DNA, and reproduce themselves. Thus, they defy the central dogma; they can make genetic information flow in the reverse (RNA → DNA) of its normal direction. And to do this they have a special tool, an enzyme called *reverse transcriptase*. Reverse transcriptase is a DNA polymerase that uses RNA as its template.

Among these viruses are tumor viruses and HIV. They are retroviruses, which means that they have the ability to insert their genetic material into the genome of the cells that they infect. But first they must transcribe their genetic material (RNA) into the genetic material of the host's genome (DNA). DNA is normally created using only other DNA strands as a template. But retroviruses make DNA copies of their RNA genome using reverse transcriptase. These then integrate into the host cell DNA, and later instruct the cell to make additional copies of the virus.

The scientists who discovered reverse transcriptase, and with it much about retroviral function, worked in teams headed by Howard Temin and David Baltimore. They were interested in RNA viruses and approached the questions about the viruses' effect on host cell RNA synthesis as a problem in enzymology. Once they had discovered that these RNA viruses contain an enzyme within their capsid (a DNA polymerase that uses RNA as its template), it stimulated a great deal of interest and work. A correspondent of the journal *Nature* called their process "reverse transcriptase" and the name stuck. Shortly afterward, the term retrovirus came into the lexicon. This was some two decades before HIV was discovered in the 1980s. But the awareness and understanding of retroviruses allowed the scientific response to AIDS.

In the first few hours after infection of a cell by a retrovirus, reverse transcription takes place and the DNA provirus that is produced integrates with the cellular DNA of the host cell genome. Then the integrated genome synthesizes progeny virus particles. These particles can rapidly spread throughout the body, infecting billions of cells.

Understanding how HIV enters cells and how it reproduces has resulted in the design of many drugs to disrupt the life cycle of the virus. Antiretroviral treatment for HIV infection consists of drugs that work against HIV infection itself by slowing down the reproduction of HIV in the body. The term highly active antiretroviral therapy (HAART) is used to describe a combination of three or more anti-HIV drugs, of which there are four main groups. Each of these groups attacks HIV in a different way.

The first group are the nucleoside reverse transcriptase inhibitors (NRTIs). They were the first type of drug available to treat HIV infection in 1987 and are better known as nucleoside analogues or nukes. As the name says, NRTIs inhibit reverse transcriptase. The drugs slow down the production of the reverse transcriptase enzyme and make HIV less able to infect cells and duplicate itself.

The second group are the non-nucleoside reverse transcriptase inhibitors (NNRTIs). These drugs started to be approved in 1997 and are generally referred to as non-nucleosides or non-nukes. This group of drugs also stops HIV from infecting cells by interfering with the trancriptase of the virus. The non-nucleoside drugs work slightly differently from the nucleoside analogues in that they bind in a different way to the reverse transcriptase. The non-nucleoside drugs also block the duplication and the spread of HIV.

The third type of antiretrovirals is the protease inhibitor (PI) group. They were first approved in 1995. Protease inhibitors, as the name says, inhibit protease. Almost every living cell contains protease. Protease is a digestive enzyme that breaks down protein and is one of the many enzymes that HIV uses to reproduce itself. The protease in HIV attacks the long healthy chains of enzymes and proteins in the host cells and cuts them into smaller pieces. These infected smaller pieces of proteins and enzymes continue to infect new cells. The protease inhibitors take effect before the protease in HIV has the chance to break down the protein and enzymes. In this way the protease inhibitors slow down the duplication of the virus and thus prevent the infection of new cells. The NRTIs and NNRTIs have an effect only on newly infected cells. Protease inhibitors are able to slow the process of immature noninfectious viruses from becoming mature and infectious. Protease inhibitors also work in cells that have been infected for a long time, by slowing down the reproduction of the virus.

The fourth group are called fusion or entry inhibitors. These drugs have yet to be approved and are currently going through clinical trials in the United Kingdom and the United States. The surface of HIV carries proteins called gp41 and gp120, which allow HIV to attach itself to, and enter into, cells. By blocking one of these proteins, fusion inhibitors slow down the reproduction of the virus. For example, T-20, the fusion inhibitor that is closest to approval, sticks to the protein gp41. The T-20 fusion inhibitor differs from the other antiretrovirals in that it needs to be injected. T-20 is a protein and cannot be taken orally, since it would be digested in the stomach.

The treatments for HIV and AIDS, while not yet cures, have transformed a rapidly fatal disease to a chronic ailment with which many people live well for decades. But here again, we must be mindful that the direction of research, and the distribution of drugs is under the control of corporate interests. As with antibiotics, the fundamental discoveries upon which these and other drugs are based were supported also by noncommercial interests. We don't need another me-too protease inhibitor. We need a new concept treatment that will destroy the giant market for HIV and AIDS drugs.

73.

The Polymerase
Chain Reaction (PCR)

**Faster than a speeding bullet, more powerful than a
locomotive . . .**

We have seen how techniques that are developed
to meet our scientific needs can facilitate discovery. Dissection, microscopy, microbe and
tissue culture, radioimmunoassay (RIA), and
imaging technologies are all among the methods
that have captured the imagination of medical scientists and have continued to contribute to advancement. The polymerase chain reaction
(PCR) is one of these wonderful tools. Invented by Kary Mullis in 1983,
PCR has revolutionized genetic research and is now used in most
modern medical-scientific laboratories, including those concerned with
forensic medicine, genetic disease diagnosis, and molecular evolution.

The purpose of a PCR is to make millions of copies of a gene.
This is essential to provide enough of a starting template for
sequencing the gene (determining the sequence of nucleotides—see
advances 27 and 28).

In the natural process that takes place within dividing cells, DNA
replicates through a series of enzyme-mediated reactions. (An enzyme
is a specific protein that controls a specific chemical reaction.) The
result is a copy of the entire genome of the species. Enzymes first

unwind the double helix of the DNA molecule, producing two single strands of DNA. This first step is called "denaturation." In the second step, an RNA polymerase enzyme synthesizes a short stretch of RNA complementary to one of the DNA strands at the beginning of the sequence that is to be copied. This "DNA/RNA heteroduplex" serves as a marker, or a "priming site," to instruct DNA polymerase where to attach and begin producing the complementary DNA.

In PCR, which takes place in the laboratory, a gene (a stretch of double-stranded DNA) is taken apart ("denatured") by exposing it to heat (94° C). Once the double-stranded DNA is separated into single strands, synthetic sequences (twenty to thirty nucleotides) of single-stranded DNA are used as primers. Two different DNA primers are used to bracket the target region to be copied (amplified). One primer is complementary to one strand of DNA at the beginning of the target region; the second primer is complementary to the other DNA strand at the end of the target region.

The PCR reaction takes place in a test tube in which there is a buffer solution containing a small amount of the target region of DNA (the gene to be amplified), a heat-stable DNA polymerase enzyme, the short oligonucleotide primers, the four deoxynucleotide building blocks of DNA (A, T, C, G), and the cofactor $MgCl_2$. The test tube is placed in an automated thermal cycler and taken through the following temperature cycles:

- several minutes at 94 to 96°C, during which the target DNA is denatured into single strands
- several minutes at 50 to 65°C, during which the primers hybridize or "anneal" (by way of hydrogen bonding) to their complementary sequences on either side of the target sequence, and
- several minutes at 72°C, during which the DNA polymerase binds and extends a complimentary strand from each primer.

The DNA sequence between the primers doubles with each cycle. Thus, after thirty cycles, or within a few minutes, the test tube contains about one billion copies of the gene. Now the genetic material can be examined, sequenced, identified, and otherwise worked with.

74.

The Human Genome Project

The technological quest to determine the order of the three billion units of DNA that make up the genetic programming of human cells—known as the human genome—began in 1990 and was projected to take fifteen years. But in September 1998 the timetable was accelerated dramatically.

Feeling intense pressure from a commercial venture, the National Human Genome Research Institute, a division of the National Institutes of Health, moved up the target date for completion by two years to 2003 and moved up the completion of 50 percent to 2001. The commercial venture challenging the NIH's formerly exclusive obsession with human genome sequencing was being spearheaded by Celera, a commercial company formed by the scientific instrument maker Perkin-Elmer and Dr. J. Craig Venter, a biologist who pioneered the sequencing of bacterial genomes. Dr. Venter announced in early 1998 that he would begin sequencing the human genome the next year, completing it by 2001. But his methods and even his ultimate goal were widely criticized by biologists who said that even if his risky methods worked, they would not produce a DNA sequence as complete and accurate as the NIH approach.[100]

Detailed knowledge of the human genome will be essential for the understanding and repair of genetic diseases. The vast DNA sequences that do not code for protein synthesis but rather control gene expression will be understood and possibly manipulated for beneficial purposes. In short, while the practical utility of the Human Genome Project is in its infancy, it someday may achieve the status of one of the very greatest human achievements.

The fundamental difference between the two groups' methodologies is that the NIH, using a mapping method, was attempting to break DNA into fragments and locate the position of each fragment on the chromosome. Dr. Venter, on the other hand, was trying to sequence millions of DNA fragments and piece them together through their overlaps, without the creation of an initial map.

In February 2001 the battle ended when the two competing groups announced the successful mapping of 85 percent of the human genome together at a ceremony at the White House. Then, on April 14, 2003, the publicly funded NIH project announced the completion of the human genome map. "We have before us the instruction set that carries each of us from the one-cell egg through adulthood to the grave," Dr. Robert Waterston, a genome sequencer who worked on the project, said at a news conference at the NIH. The Human Genome Project has determined a sequence of more than three billion units of DNA of the human genome that accounts for 99.99 percent accurate completion. They accomplished their goal two years earlier than initially expected as a result of the intense competition, and the project came in $300 million under budget ($2.7 billion).[101]

But the successful sequencing has not ended the controversies. Until April 2005, Celera had refused to release its information into the public domain, instead charging millions of dollars for it. The company had publicly stated its desire to become the "Bloomberg of biology"[102] referring to Michael Bloomberg's multimillion-dollar finance information empire. But with the public project offering much of the same information for free, Celera was unable to turn a profit. It has abandoned the effort and will discontinue the service, which it called the Celera Discovery System, by donating the thirty billion base pairs to a federally run database. Dr. Venter said in a statement after the announcement that "[m]oving the Celera data into the public domain is something I have been strongly in favor of, and I feel it sets a good precedent for companies who are sitting on gene and genome data sets that have little or no commercial value, but would be of great benefit to the scientific community."[103]

75.

Gene Therapy

The best laid schemes o' mice an' men gang aft a-gley.

—Robert Burns, from "To a Mouse"

Replacing defective genes with healthy ones, and thereby fixing the molecular cause of disease instead of simply treating its symptoms, is a powerful approach that is in its infancy. But it was born with defects that must be understood and corrected. Despite many efforts, gene therapy appears to have cured only about a dozen patients, all of them children in Europe who were born with a severe immune system disorder. Three of those children have developed cancer despite the treatment and one recently died, leaving even the field's active proponents discouraged.

Gene therapy is, of course, an experimental treatment. The method involves infusions of mouse viruses engineered to carry the immune system gene that these patients lack (see advance 42 and 43). The viruses infect the patients' immune system cells to deliver the needed gene. But the viruses sometimes disturb healthy genes—including genes that, when disrupted, can cause cancer.

In March 2005 the Food and Drug Administration (FDA) suspended US gene therapy experiments after learning that the third child who underwent treatment in France had developed cancer. In addition, a monkey has recently died of cancer caused by a gene therapy experiment six years ago.

The action by the FDA was judicious and timely. It is consistent with Hippocrates' first principle of "do no harm." Gene therapists are now focused on these problems. They are conducting more research to work toward safer studies in the future. Although much more work needs to be done, I believe that this approach will fulfill its great promise.

76.

The Electrocardiogram

The electrocardiogram is an indispensable tool for evaluating the heart. Its introduction, in the early years of the twentieth century, improved the diagnosis and treatment of heart diseases and gave us the image of the pulse's tracing as an icon in our collective consciousness. It is immediately needed, just after a brief history and physical examination, in the emergency evaluation of chest pain and irregular heartbeat. Its findings direct our response, and it is often a lifesaver. There is nothing that makes an apprentice healer feel growing accomplishment and security more than learning to read the language of electrocardiograms. Today, they are routinely read by built-in computers, and in the early 1960s it was felt that within a few years computer reading would be more reliable than human readings. We're not there yet.

An electrocardiogram (often abbreviated as EKG from the German *Elektrokardiogram*) is a graphic record of the electrical activity of a patient's heartbeat. Each heartbeat is powered by an electrical impulse traveling through the heart. An EKG describes the pace of the heartbeat, the power (amplitude) of the electrical charges, and the direction of their travels.

The first person to accurately measure and record the heart's electrical activity, thereby inventing the EKG, was the Dutch physiologist Willem Einthoven. In 1903 he created a rather simple string-galvanometer that was able to measure the electrical potential of a beating heart. Electrodes were attached to the limbs of the patient, and as the string deflected, it obstructed a beam of light and the photographic paper recorded the shadow. Einthoven was awarded the Nobel Prize in 1924 for this discovery, which immediately changed the effectiveness of cardiac care. Einthoven described five separate heart waves: P, Q, R, S, and T.

The P wave represents the electrical signature of the current that contracts the atria (upper chambers of the heart). The left and right atria contract simultaneously, and an irregular or absent P wave may indicate arrhythmia.

The Q, R, and S complex of waves represents the electrical charges moving through the left and right ventricles, which are much stronger than those from the atria.

The T wave is a representation of the electrical repolarization of the ventricles. The T wave is much weaker than the other four, and if one thinks of the cardiac muscle cells as loaded springs, then the T wave is the spring resetting itself in anticipation of the next electrical impulse.

The shape of the waves (peaked, flat, notched, inverted, coveplain), the intervals between them (P-R interval, Q-T interval), can describe anatomic and functional impediments to the flow of electricity through the heart.

The EKG can diagnose most heart attacks, predict their location in the heart as well as their severity, and monitor their progress. It is essential for the precise diagnosis of irregular heartbeats. Among modern inventions, only the x-ray machine has had a comparable impact on medicine.

77.

Mitochondrial Function and Disease

L eeuwenhoek's "craving after knowledge" led him to discover cells of all kinds (see advance 4). He described animal and plant cells (eukaryotes), and bacteria (prokaryotes). He even described some of the fine structures within cells. To reiterate, the impact of his discoveries on biology and all the medical sciences cannot be overstated. Microscopes were irrevocably turned to the study of cell biology.

Life arose on Earth about four billion years ago. The first cells to evolve were prokaryotic cells, simple single-celled organisms that lack a nuclear membrane. Bacteria are the best-known and most-studied prokaryotes. They contain DNA, just as we do, but it is not confined within a membrane, so there is no distinct nuclear structure, and there are other ways in which our cells differ. The recent discovery of a second group of prokaryotes, the archaea, has disclosed a third cellular form of life, giving us new insights into life's origins. Archaea are more closely related to eukarya than are the bacteria.

The organelles (tiny organlike structures within cells) of eukary-

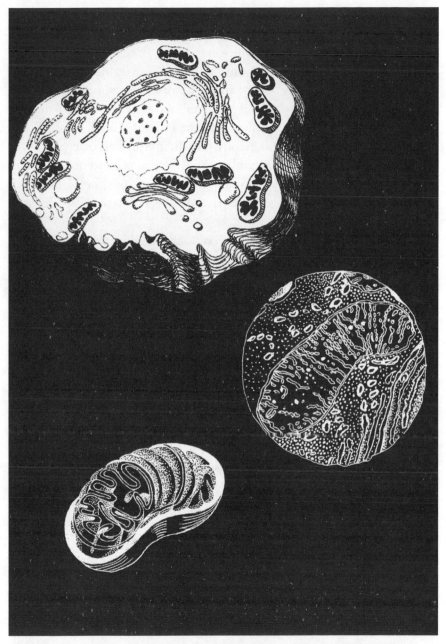

Fig. 47. Drawings of mitochondria shown within a cell (*top*), a closer view (*middle*), and a single mitochondria (*below*).

otes (mitochondria in animal cells and chloroplasts in plants) are thought to be remnants of bacteria that invaded, or were captured by, primitive eukaryotes in the evolutionary past. Prokaryotes have many symbiotic relationships with plants and animals. By definition, these associations benefit both organisms. The diseases caused by bacteria are the exception, not the rule. The metabolic activities of prokaryotes in soil are the basis of soil fertility, and the bulk of Earth's atmospheric oxygen may have been produced by free-living prokaryotic cells.

But understanding endosymbiosis, where a prokaryote becomes integrated into the fundamental structure of our cells and our genome—becoming an essential part of us—is, at the very least, a challenge.

Mitochondria are the engines of our cellular metabolism, the place where cellular respiration takes place. Breathing in and out to take in oxygen and excrete carbon dioxide is the obvious respiration with which we are all familiar. But there is another aspect of respiration that is hidden in our cells. This is the oxidation of food chemicals such as glucose to produce energy, and it is the job of the mitochondria. Our cells have evolved to require oxygen in order for energy to be produced. This energy is stored in the chemical bonds of adenine triphosphate (ATP). The process through which ATP is created in the presence of oxygen takes place in the mitochondria. It is called aerobic respiration, or cellular respiration, and it normally is coupled to oxidative phosphoralation. Oxidative phosphoralation is the payoff in the oxidation of glucose to carbon dioxide and water. It is the step where the energy released is trapped in the high-energy bond that forms when inorganic phosphorus is combined with adenine diphosphate (ADP) to produce ATP. The energy is then used for all the energy-consuming activities of the cell.

The process occurs in two phases: glycolysis, the breakdown of glucose to pyruvic acid, and then the complete oxidation of pyruvic acid to carbon dioxide and water.

The details of cellular respiration—the enzymes involved, their locations in the mitochondria, the chemical intermediates, the quantitative aspects of ATP production and utilization—are fundamental to cell biology, and therefore to medicine. The elucidation of these wonders, by people like Carl and Gerty Cori, are surely among the greatest medical advances. And the description of what happens to people when mitochondria are dysfunctional was similarly ground-breaking, opening the field of "mitochondrial medicine."

In 1962 Rolf Luft and his colleagues in Stockholm described a patient in whom the energy release by cellular respiration was not coupled with the phosphoralation of ADP.[104] Insufficient ATP was formed and the energy was being wasted as heat. The patient was a woman who had suffered for years with weight loss, muscular wasting, weakness, sweating, and excessive drinking and eating. In elegant studies of mitochondria isolated from her muscles, the group showed "a loosely coupled state of the oxidative phosphorylation in the patient's mitochondria." That is, the energy released was inefficiently trapped in the high-energy bond that forms when inorganic phosphorus is combined with ADP. So less ATP was being produced. Electron microcopy (advance 30) showed that at a magnification of 65,000X to 135,000X, the mitochondria looked quite abnormal.

We now know that mitochondrial disease is not rare, and the concepts and methods used to establish this advance would represent many of our one hundred greatest advances.

78.

Freud and the Realization of the Unconscious Mind

Know thyself.

—Inscription on the Oracle of Apollo at
Delphi, Greece (sixth century BCE)

I have found, in my own case too, [the phenomenon of] being in love with my mother and jealous of my father, and I now consider it a universal event in early childhood, even if not so early as in children who have been made hysterical.

—Sigmund Freud (1856–1939)

Sigmund Freud was a man of culture, of linguistic virtuosity, of elegance, and of power in writing, who could inform, entertain, and galvanize a sitting room, a lecture hall, a country, or the whole world through the sheer originality, profundity, and courage of his thinking. He uncovered the unconscious mind—with our deepest desires, our greatest fears, our unthinkable wishes—and for this he was both worshiped and despised, often by the same people at the same time. And he made most of his discoveries by simply looking inward with an unflinching eye. Much of civilization has been affected by his ideas about mental health and disease.

Freud's method of inquiry and treatment, psychoanalysis, while still taught and practiced, is antithetical to current directions in healthcare. It relies on time and the painstaking examination of the patient's life, especially in early childhood. It delves into memories, feelings, dreams, thoughts, reactions, and associations. We are most familiar with the method through his many publications, including fragments of his pivotal cases including Elizabeth von R., Dora, Little Hans, the Man with the Rats, the Man with the Wolves, President Schreber, and others from his writings. These cases taught Freud, and the world, about the workings of the mind, as evident in the title of one of his books, *The Psychopathology of Everyday Life.* Nevertheless, the idea of a patient spending three to five hours in a psychiatrist's office every week for years is an insurance company's nightmare and is inconsistent with the pace of contemporary life. Today, Freudian psychoanalysis is essentially limited to advanced students of analytic technique. But, in our opinion, all other methods of psychotherapy, including eclectic mixtures of Jungian therapy, cognitive therapy, even various group therapies, are based upon Freudian insights. And whatever one's beliefs about the efficacy of the talking cure (as it was often called), psychoanalysis was the method through which Freud discovered the Oedipus complex; the technique of free association; his ideas about the role of sexuality in human development; neuroses, hysteria, and phobias; the nature of dreams and their interpretation; obsessive behavior; jokes and humor; and paranoia. Though many of his views may no longer be in vogue, he brought revolutionary insight into the vast array of human behaviors and psychological afflictions.

He was, to say the least, influential and controversial. Totalitarians despised him. In 1933 Adolf Hitler held a special book burning in Berlin dedicated to the works of Sigmund Freud. Stalinists were at least deeply wary, as wary as he was of them:

> The enthusiasm with which the mob follow the . . . lead at present, so long as the new order is incomplete and threatened from outside, gives no guarantee for the future, when it will be fully established and no longer in danger. In exactly the same way as religion, [authoritarianism] is obliged to compensate its believers for the sufferings and deprivations of the present life by promising them a

Fig. 48. Hypnosis seeks contact with the unconscious mind.

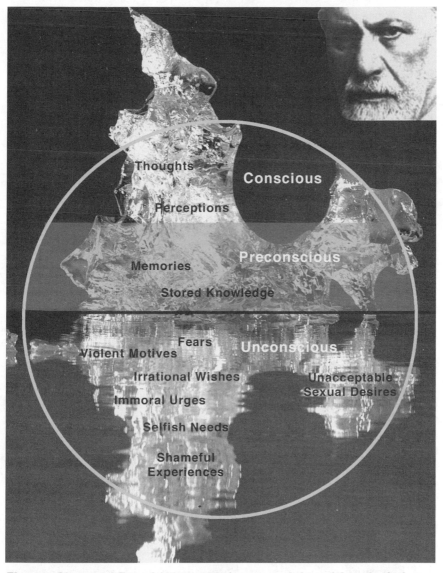

Fig. 49. Sigmund Freud (1856–1939), *upper right,* **with a depiction of the mind using the metaphor of an iceberg. Consciousness and preconsciousness are shown above the surface with the unconscious mind represented below.**

better life hereafter. . . . It is true that this paradise is to be in this world; it will be established on earth, and will be inaugurated within a measurable time. But let us remember that the Jews, whose religion knows nothing of a life beyond the grave, also expected the coming of the Messiah here on earth, and that the Christian Middle Ages constantly believed that the Kingdom of God was at hand.[105]

Freud argued that the imposition of society upon human beings, just as the introduction of children into social groups, requires the repression of fundamental desires and instincts. Moreover, he believed that civilization itself is a substitute for our ancient drives and behaviors. In his *Introductory Lectures on Psycho-Analysis* (first lecture), he says, "we believe that civilization is to a large extent being constantly created anew, since each individual who makes a fresh entry into human society repeats this sacrifice of instinctual satisfaction for the benefit of the whole community." The "primitive impulses," among which sexuality is the most powerful, are sublimated or "diverted" toward other goals that are "socially higher and no longer sexual." These "atavistic" urges are repressed in order for us to function within the group. But Freud felt that sexual impulses are so powerful that they often "return" from unconsciousness to influence conscious behavior.[106]

"The prehistory into which the dream-work leads us back is of two kinds—on the one hand, into the individual's prehistory, his childhood, and on the other, in so far as each individual somehow recapitulates in an abbreviated form the entire development of the human race, into phylogenetic prehistory too . . . symbolic connections, which the individual has never acquired by learning, may justly claim to be regarded as a phylogenetic heritage."[107] In other words, as in the physical realm of embryology where "ontogeny recapitulates phylogeny," or the development of each individual briefly goes through the evolution of our species, so it is also with the mind.

The repressed content of the mind normally exists in a state of balance, or equilibrium, with consciousness. We get glimpses into its presence in dreams and seemingly insignificant mistakes, called Freudian slips (or as he called them, "parapraxes"), which for him were clues to the mysterious and secret activities of the unconscious mind.

Freud's model of the mind, expressed in terms of three overlapping and interacting areas called the superego, the ego, and the id—each of which contributing to preconsciousness, repressed consciousness, and unconsciousness—has become deeply ingrained in human culture. Very briefly, the id is the storehouse of sexual desire (libido), from which the ego tries to separate itself by means of repression. So the id expresses sexuality in terms of "screen-memories" that the ego can then deal with in a conscious fashion. As Freud put it in *Ego and the Id*, "the ego has the task of bringing the influence of the external world to bear upon the id and its tendencies, and endeavors to substitute the reality-principle for the pleasure-principle which reigns supreme in the id." So the ego is concerned with perceptions in the external world of reason and considered judgment, while the id is inclined toward instincts and passions. But the ego is enmeshed with the id and can never completely drive it out. The superego is that set of feelings and ideas which manifest as conscience and feelings of guilt. Freud felt that the superego developed as a response to the Oedipus complex and is the internalization of the limits and prohibitions imposed by the father.

This model of the psyche was radically different from anything ever proposed. Implicit in the model was the notion that the difference between the sane and the mentally ill is only a matter of degree: "[I]f you take up a theoretical point of view . . . , you may quite well say that we are all ill—if you look at the matter from a theoretical point of view and ignore this question of degree you can very well say that we are all ill, that is, neurotic—since the preconditions for the formation of symptoms can also be observed in normal people."[108]

Though Freud's model of the mind is not taken literally, it has provided us with a new way to talk about the deep and seemingly inexplicable thoughts that make up our mind and result in our actions. Freud was primarily responsible for bringing the mentally ill up from abandonment and making psychiatry a specialized branch of medicine. It is because of the unique complexity of the human brain that psychiatric categorization, description, and precise diagnosis is so difficult. Is there one schizophrenia, or are there five hundred? Do we need yet another *Diagnostic and Statistical Manual* (*DSM-V*) for psychiatric disease? Progress is difficult; we need a breakthrough, but most patients can be helped, and Freud brought mental illness out of dark shadows and into a reasonable confusion.

Freud's realization of the unconscious mind is only part of his great contribution to medicine and culture, and it is a measure of the man that the mention of his name often stimulates lively discussion, if not strenuous argument.

79.

Group and Family Therapy

My routines come out of total unhappiness. My audiences are my group therapy.

—Joan Rivers (b. 1935), US comedienne

T he tradition within which psychotherapy took shape and began to grow started with that of the couch. The patient would relax and speak, with the psychiatrist seated in a chair, slightly behind, listening. For much of its short history as a scientific field, the idea of group therapy was anathema. But psychiatry is brave and young, and perhaps more than other medical disciplines, it has evolved with respect to the playing field and the ground rules. Gone is the couch, at least as the universal symbol of the psychotherapeutic experience, and so, too, is the rule of one therapist, one patient. And while individual psychotherapy is still practiced, there is an increasing trend toward various group therapy[109] settings, among which family therapy is but one example.

Broadly described as the "group helping field," there are now many types of clinical groups offering help; these include therapeutic groups, human relations and training groups, as well as mutual-help and self-help groups. So we have gone from psychiatrist as father figure to a vast array of therapeutic relationships, mostly within my professional lifetime. And there are many theoretical models of group interventions with varying approaches to group treatment theory, practice, and research. Group interventions in the fields of physical, sexual, and substance abuse, chronic illness, and trauma—as in the aftermath of disasters like the September 11, 2001, terrorist attacks or the 2005 tsunami—are widespread. This is due to the proven effectiveness of the approach, as well as the obvious possibilities for reducing cost to payers. Interestingly, the "managed care revolution" in mental health has paradoxically been an obstacle to group and family therapy, and all other forms of psychotherapy.

The "revolution" has shaken the very foundations of the prospering group psychotherapy movement. Pressure from third-party payers for greater accountability and for cost containment has come to threaten the traditional autonomy and dedication to quality of care by group therapy practitioners. Professional organizations have begun to meet these challenges by educational endeavors directed at insurers, legislators, and the public. Current training programs and literature for clinicians have focused on the need for a new "business orientation" and on less costly group treatment measures (i.e., short-term and combined therapy).

In these and other ways, corporate medicine is shaping the mental health field, and group and family therapy is in decline. For more than thirty years, most counties have provided mental health services to those who cannot afford them. Over the last several years, counties have outsourced an increasing percent of their mental health services to private providers with the goal of offering the "more cost-efficient services." In many states, the remaining county mental health services are in the last phase of "managed competition," through which county providers bid against private mental healthcare providers for county contracts, with the lowest bidder winning.

Managed competition has revealed many aspects of privately provided mental health services, the most disconcerting of which is the fact that county residents will lack several critical mental health safety nets if these services are turned over to private companies. An example of the loss of a safety net within a county is related to the lack of emergency psychiatric services within regions under privatization. Also, private providers have refused to bear the financial burden of providing psychiatric medications when they became too costly to dispense.

Neglect, abandonment, and punishment are not ancient approaches to the mentally ill. While the mad houses, asylums, and state hospital warehouses are mostly gone, a very large fraction of the mentally ill can be found in prisions, among the homeless, and otherwise untreated. The tremendous advances in understanding the biochemical and developmental aspects of mental illness notwithstanding, there is still resistance to the medical model in mental health. Even the existence of common afflictions are often denied, like post-traumatic stress syndrome (PTSS). The "moralistic" approach, with allusions to malingering and admonitions to "pull yourself together," is still pervasive. But mental health professionals have been innovative. And while they are currently reeling backward, they are trying to make a stand in this most complex area of medical science.

80.

Twelve-Step Programs and the Treatment of Addictive Behavior

Taken in moderation, alcohol may have certain health benefits, but in excess it regularly causes disastrous behaviors, which may lead to traumatic injury and to tissue damage resulting in multiple organ failure and death. The acute, or sudden, ingestion of excessive amounts causes rapidly rising blood alcohol concentrations, which, if sufficiently high, can cause coma with high mortality rates. But it is the chronic effects of addiction that brings about the liver, brain, nerve, pancreatic, and other organ damage that has been the source of so much misery throughout the ages. Clearly, alcohol is a source of comfort and cheer for many and of tragedy for others who become addicted. Just as clearly, addiction is very difficult to treat, and while a few can simply "kick the habit," most treatment approaches have failed. Amid so much suffering and failure, a successful treatment, however controversial or even objectionable, represents a great advance. And then along came Bill W. and Dr. Bob.

Bill W. and Dr. Bob were drunks in every sense of the word. Bill Wilson, a hard-drinking New York stockbroker, and Bob Smith, an Ohio-based physician and local minister, met in 1934 through a mutual friend. Bill spent prohibition drinking three bottles of "bathtub" gin a day, while Dr. Bob was boozing his way through three failed marriages.

Within a year, the two men had not only conquered their own addiction but set out to share their successes through a revolutionary combination of spiritual submission, meditation, and group therapy, now acknowledged by many as the most successful addiction-control program in human history.

Alcoholics Anonymous, their 1939 best-seller known by the AA faithful simply as the "Big Book," communicated the core ideas that would soon evolve into the world-famous 12-step program. Today, after selling over twenty million copies, the organization universally recognized by the acronym AA boasts over two million members, with nearly 100,800 groups meeting in 150 countries. These groups, which meet at least once a week, have always been absolutely critical to the AA recovery process. "I am John Doe, and I am an alcoholic," is the opening and mandatory introduction every member must perform as a first step toward his/her public testimonial. These now-famous testimonials, the first of which were from founders Bill and Bob and have appeared in every edition of their book, are used as both a public admission and a lifelong acceptance of vulnerability to relapse. AA members insist, and independent research supports,[110] that the realization of addiction as a disease and not simply a temporary condition is critical to permanent recovery.

Yet, despite its well-documented successes, controversy has followed the program from the outset. Based loosely on the principals of the Oxford Group, a spiritual and decidedly Protestant organization whose founder, Frank Buchman, had been accused of having pro-Hitler leanings and authoritarian ambitions,[111] AA has been criticized throughout its history for promoting an almost brainwashed and cult-like following. Critics point to its emphasis on the powerlessness of the individual and the program's reliance on God as savior. A careful analysis of the twelve steps in their entirety both supports this view and simultaneously illustrates a conscious effort on the part of its authors to obscure that same argument:

1. Admit that you are powerless over alcohol—that your life has become unmanageable.

2. Come to believe that a Power greater than yourself can restore you to sanity.

3. Make a decision to turn your will and your life over to the care of God a*s you understand Him*.

4. Make a searching and fearless moral inventory of yourself.

5. Admit to God, and to yourself, and to another human being the exact nature of your wrongs.

6. Be entirely ready to have God remove all of these defects of character.

7. Humbly ask Him to remove your shortcomings.

8. Make a list of persons that you have harmed, and become willing to make amends to them all.

9. Make direct amends to such people wherever possible, except when to do so would injure them or others.

10. Continue to take personal inventory and when you are wrong promptly admit it.

11. Seek through prayer and meditation to improve your conscious contact with God, *as you understand Him*, praying only for knowledge of His will for you and the power to carry that out.

12. Having had a spiritual awakening as a result of these steps, try to carry this message to alcoholics, and practice these principals in all of your affairs.

Clearly, eight of the twelve steps made famous by AA's founders include either a direct or an obvious reference to God. Yet Bill Wilson, who was never a religious man, made sure that these references were left open to personal interpretation, by inserting the inclusive phrase: *as you understand Him*.

There continue to be, however, many people who feel AA's twelve steps exclude thousands of nonspiritual people and that the notion of powerlessness over addiction—that addicted people are always recovering as opposed to cured, and that only a higher power can protect against relapses—is inherently destructive. Groups such as the Secular Organization for Sobriety, which bills itself as "the world's largest non-religious alternative to the 12-step," and another California-based secular organization called Rational Recovery, whose founder has publicly stated that "AA is a religion, but it is a false religion which tells everyone they must deny free will," serve to keep this debate alive.[112]

Regardless of the surrounding controversy, AA and its twelve steps have moved well beyond just alcohol addiction. There are now over two hundred types of "12-step" recovery programs in the United States, such as Narcotics Anonymous, Overeaters Anonymous, and even Shoppers Anonymous. And the National Institutes of Health recognizes three "scientifically based approaches to drug addiction treatment," all of which borrow significantly from Bill W. and Dr. Bob's seventy-year-old formula.

81.

Methadone Maintenance Treatment

The early 1960s saw a virtual epidemic of heroin and morphine abuse in America. Addicts of these potent and highly addictive drugs very often became serial criminals and overdose victims. Unfortunately, the only accepted treatments at the time—detoxification centers, lobotomies, insulin shock, and jail—were all experiencing extremely low rates of success. At the same time, however, doctors from New York City to Lexington, Kentucky, were experimenting with highly controversial opiate inhibitors (receptor antagonists) such as methadone. It soon became clear that methadone maintenance was the one and only treatment option achieving any positive results. But it remained highly controversial.

Methadone, like heroin and morphine, is an addictive and controlled substance. However, unlike those drugs methadone does not produce feelings of euphoria and, most important, it has the ability to bind to the "opiate" receptors (heroin and morphine are opium derivatives) reducing the craving for other opiates. This phenomenon makes methadone, in technical terms, a competitive antagonist. In Paul Ehrlich's terms, a "magic bullet" (scc advance 22).

Taken once a day, methadone can suppress narcotic withdrawal and completely eliminate the positive effects of opiate drugs if and when they are taken. As a result, methadone patients do not experience the extreme highs and lows that varying levels of opiates in the blood produce. In other words, while the patient does remain physically dependent on methadone, he is not overcome by the uncontrolled, compulsive, and disruptive behavior that defines heroin addicts. Also, methadone produces a significantly slower withdrawal process and therefore it is possible to maintain a methadone addiction without harsh and dangerous side effects.

A majority of patients require 80 to 120 mg/d of methadone or more to combat cravings and the National Institute on Drug Abuse found that, among outpatients receiving methadone maintenance treatment (MMT), weekly heroin use decreased by 69 percent. Nevertheless, MMT is still controversial, and treatment facilities have trouble getting the proper medication despite the fact that methadone maintenance has been found to be medically safe and nonsedating. Reviews issued by the Institute of Medicine and the National Institutes of Health claim that methadone maintenance coupled with social services is the most effective treatment for narcotic addiction. These agencies recommend reducing governmental regulation to facilitate patients' access to treatment and recommend that methadone maintenance be available to persons under legal supervision, such as probationers, parolees, and the incarcerated. Yet a debilitating stigma and bias continues to handicap these programs and its patients, compromising the effectiveness of services.[113]

Treatment of Schizophrenia

> but each of us
> has his own kingdom of pains
> and has not yet found them all
> and is sailing in search of them day and night
> infallible undisputed unresting
> filled with a dumb use
> and its time
> like a finger in a world without hands
>
> —W. S. Merwin, from "Beggars and Kings"

> What's madness but nobility of soul
> At odds with circumstance? The day's on fire!
> I know the purity of pure despair,
> My shadow pinned against a sweating wall,
> That place among the rocks—is it a cave,
> Or winding path? The edge is what I have.
>
> —Theodore Roethke, from "In a Dark Time"

Major mental illness has confused, frightened, and angered people throughout recorded history. The mentally ill have been subject to treatments that punished them and increased their suffering. Only recently has some understanding emerged and, more recently, have effective treatments been introduced. The following information is culled from the National Institute of Mental Health, NIH Publication No. 02-3517 (1999, 2002).

WHAT IS SCHIZOPHRENIA?

Schizophrenia is a chronic, severe, and disabling brain disease. Approximately 1 percent of the population develops schizophrenia during its lifetime—more than two million Americans suffer from the illness in a given year. Although schizophrenia affects men and women with equal frequency, the disorder often appears earlier in men, usually in the late teens or early twenties, than in women, who are generally affected in the twenties to early thirties. People with schizophrenia often suffer terrifying symptoms such as hearing internal voices not heard by others or believing that other people are reading their minds, controlling their thoughts, or plotting to harm them. These symptoms may leave them fearful and withdrawn. Their speech and behavior can be so disorganized that they may be incomprehensible or frightening to others. Available treatments can relieve many symptoms, but most people with schizophrenia continue to suffer some symptoms throughout their lives; it has been estimated that no more than one in five individuals recovers completely.

This is a time of hope for people with schizophrenia and their families. Research is gradually leading to new and safer medications and unraveling the complex causes of the disease. Scientists are using many approaches from the study of molecular genetics to the study of populations to learn about schizophrenia. Methods of imaging the brain's structure and function hold the promise of new insights into the disorder.

Schizophrenia as an Illness

Schizophrenia is found all over the world. The severity of the symptoms and the long-lasting, chronic pattern of schizophrenia often cause a high degree of disability. Medications and other treatments for schizophrenia, when used regularly and as prescribed, can help reduce and control the distressing symptoms of the illness. However,

Fig. 50. Metaphor for multiple personalities within one person.

some people are not greatly helped by available treatments or may prematurely discontinue treatment because of unpleasant side effects or other reasons. Even when treatment is effective, persisting consequences of the illness—lost opportunities, stigma, residual symptoms, and medication side effects—may be very troubling.

Developmental neurobiologists funded by the National Institute of Mental Health (NIMH) have found that schizophrenia may be a developmental disorder that occurs when neurons form inappropriate connections during fetal development. These errors may lie dormant until puberty, when changes in the brain that occur normally during this critical stage of maturation interact adversely with the faulty connections. This research has spurred efforts to identify prenatal factors that may have some bearing on the apparent developmental abnormality.

In other studies, investigators using brain-imaging techniques have found evidence of early biochemical changes that may precede the onset of disease symptoms, prompting examination of the neural circuits that are most likely to be involved in producing those symptoms. Meanwhile, scientists working at the molecular level are exploring the genetic basis for abnormalities in brain development and in the neurotransmitter systems regulating brain function.

HOW IS IT TREATED?

Since schizophrenia may not be a single condition and its causes are not yet known, current treatment methods are based on both clinical research and experience. These approaches are chosen on the basis of their ability to reduce the symptoms of schizophrenia and to lessen the chances that symptoms will return.

What about Medications?

Antipsychotic medications have been available since the mid-1950s. They have greatly improved the outlook for individual patients. These medications reduce the psychotic symptoms of schizophrenia and usually allow the patient to function more effectively and appropriately. Antipsychotic drugs are the best treatment now available, but they do not "cure" schizophrenia or ensure that there will be no further psychotic episodes. The choice and dosage of medication can be made only by a qualified physician who is well trained in the medical treatment of mental disorders. The dosage of medication is individualized for each patient, since people may vary a great deal in the amount of drug needed to reduce symptoms without producing troublesome side effects.

The large majority of people with schizophrenia show substantial improvement when treated with antipsychotic drugs. Some patients, however, are not helped very much by the medications, and a few do not seem to need them. It is difficult to predict which patients will fall into these two groups and to distinguish them from the large majority of patients who do benefit from treatment with antipsychotic drugs.

A number of new antipsychotic drugs (the so-called atypical antipsychotics) have been introduced since 1990. The first of these, clozapine (Clozaril), has been shown to be more effective than other antipsychotics, although the possibility of severe side effects—in particular, a condition called agranulocytosis (loss of the white blood cells that fight infection)—requires that patients be monitored with blood tests every one or two weeks. Newer antipsychotic drugs, such as risperidone (Risperdal) and olanzapine (Zyprexa), are safer than the older drugs or clozapine, and they also may be better tolerated. They may or may not treat the illness as well as clozapine, however. Several additional antipsychotics are currently under development.

Antipsychotic drugs are often very effective in treating certain symptoms of schizophrenia, particularly hallucinations and delusions; unfortunately, the drugs may not be as helpful with other symptoms, such as reduced motivation and emotional expressiveness.

Patients and families sometimes become worried about the antipsychotic medications used to treat schizophrenia. In addition to concern about side effects, they may worry that such drugs could lead to addiction. However, antipsychotic medications do not produce a "high" (euphoria) or addictive behavior in people who take them.

Another misconception about antipsychotic drugs is that they act as a kind of mind control, or a "chemical straitjacket." Antipsychotic drugs used at the appropriate dosage do not "knock out" people or take away their free will. While these medications can be sedating, and while this effect can be useful when treatment is initiated particularly if an individual is quite agitated, the utility of the drugs is not due to sedation but to their ability to diminish the hallucinations, agitation, confusion, and delusions of a psychotic episode. Thus, antipsychotic medications should eventually help an individual with schizophrenia to deal with the world more rationally.

Antipsychotic drugs, like virtually all medications, have unwanted effects along with their beneficial effects. During the early phases of drug treatment, patients may be troubled by side effects such as drowsiness, restlessness, muscle spasms, tremor, dry mouth, or blurring of vision. Most of these can be corrected by lowering the dosage or by using other medications. Different patients have different treatment responses and side effects to various antipsychotic drugs. A patient may do better with one drug than another.

The long-term side effects of antipsychotic drugs may pose a considerably more serious problem. Tardive dyskinesia (TD) is a disorder characterized by involuntary movements most often affecting the mouth, lips, and tongue, and sometimes the trunk or other parts of the body such as the arms and the legs. It occurs in about 15 to 20 percent of patients who have been receiving the older, "typical" antipsychotic drugs for many years, but TD can also develop in patients who have been treated with these drugs for shorter periods of time. In most cases, the symptoms of TD are mild, and the patient may be unaware of the movements.

While psychosocial approaches have limited value for acutely psychotic patients (those who are out of touch with reality or have prominent hallucinations or delusions), they may be useful for patients with less severe symptoms or for patients whose psychotic symptoms are under control. Numerous forms of psychosocial therapy are available for people with schizophrenia, and most focus on improving the patient's social functioning—whether in the hospital or community, at home, or on the job.

While the future for schizophrenics should continue to improve, there are still major problems. How are they to obtain and maintain their medications and other treatments? And a related issue: the legacy of abuse and the asylum continue as the ranks of the homeless and prison inmates swell with the mentally ill.

83.

Treatment of Depression

**I been down so long,
It seems like up to me.**

—Furry Lewis (1893–1981)

The above lyrics provide a window into clinical depression as something more than just a blue mood. This short refrain expresses a number of the clinical criteria required for the diagnosis: the mood of sadness and emptiness, the element of chronicity, and the feeling of hopelessness. Clinical depression is not just a low period in response to tragedy or hardship. It is a common affliction suffered by millions of people and may be associated with great personal costs. Literature and history make it clear that depression and changing attitudes toward it have been with us from antiquity.

Depression is second only to hypertension as the most common chronic condition encountered in general medical practice. In each case the diagnosis should be made by a qualified expert, and treatment should be initiated and supervised by experienced professionals. And this is all the more important because major depressive illnesses are increasingly seen as chemical imbalances in the brain for which potent chemotherapeutic agents are available. Taken alone, or in combination with other medications or psychotherapy, most people can now be restored to well-being. This is certainly one of the greatest medical advances.

The problem is that, while help is available, many people are unaware of the nature of their illness and the fact that there are successful treatments. The National Institute of Mental Health has embarked on an outreach and education program that can reduce the personal and societal suffering caused by depression.

The many medications that may be used are beyond the scope of this chapter, but it is important to emphasize that they may require change and adjustment in dosage before optimal results are achieved. Nevertheless, under good supervision success can now be expected in many cases.

Depression

In any given one-year period, 9.5 percent of the population, or about 18.8 million American adults, suffer from a depressive illness. The economic cost for this disorder is high, but the cost in human suffering cannot be estimated. Depressive illnesses often interfere with normal functioning and cause pain and suffering not only to those who have a disorder but also to those who care about them. Serious depression can destroy family life as well as the life of the ill person. But much of this suffering is unnecessary.

Most people with a depressive illness do not seek treatment, although the great majority—even those whose depression is extremely severe—can be helped. Thanks to years of fruitful research, there are now medications and psychosocial therapies such as cognitive/behavioral, "talk," or interpersonal that ease the pain of depression.

Unfortunately, many people do not recognize that depression is a treatable illness. If you feel that you or someone you care about is one of the many undiagnosed depressed people in this country, the information presented here may help you take the steps that may save your own or someone else's life.

What Is a Depressive Disorder?

A depressive disorder is an illness that involves the body, mood, and thoughts. It affects the way a person eats and sleeps, the way one feels about oneself, and the way one thinks about things. A depressive disorder is not the same as a passing blue mood. It is not a sign of personal weakness or a condition that can be willed or wished away. People with a depressive illness cannot merely "pull themselves together" and get better. Without treatment, symptoms can last for weeks, months, or years. Appropriate treatment, however, can help most people who suffer from depression.

Types of Depression

Depressive disorders come in different forms, just as is the case with other illnesses such as heart disease. Here we briefly describe three of the most common types of depressive disorders. However, within these types there are variations in the number of symptoms, their severity, and persistence.

Major depression is manifested by a combination of symptoms (see symptom list) that interfere with the ability to work, study, sleep, eat, and enjoy once pleasurable activities. Such a disabling episode of depression may occur only once but more commonly occurs several times in a lifetime.

A less severe type of depression, dysthymia, involves long-term, chronic symptoms that do not disable but keep one from functioning well or from feeling good. Many people with dysthymia also experience major depressive episodes at some time in their lives.

Another type of depression is bipolar disorder, also called manic-depressive illness. Not nearly as prevalent as other forms of depressive disorders, bipolar disorder is characterized by cycling mood changes: severe highs (mania) and lows (depression). Sometimes the mood switches are dramatic and rapid, but most often they are gradual. When in the depressed cycle, an individual can have any or all of the symptoms of a depressive disorder. When in the manic cycle, the individual may be overactive, overtalkative, and have a great deal of energy. Mania often affects thinking, judgment, and social behavior in ways that cause serious problems and embarrassment. For example, the individual in a manic phase may feel elated, full of grand schemes that might range from unwise business decisions to romantic sprees. Mania, left untreated, may worsen to a psychotic state.

Symptoms of Depression and Mania

Not everyone who is depressed or manic experiences every symptom. Some people experience a few symptoms, some many. Severity of symptoms varies with individuals and also varies over time.

Depression

- Persistent sad, anxious, or "empty" mood
- Feelings of hopelessness, pessimism
- Feelings of guilt, worthlessness, helplessness
- Loss of interest or pleasure in hobbies and activities that were once enjoyed, including sex
- Decreased energy, fatigue, being "slowed down"
- Difficulty concentrating, remembering, making decisions
- Insomnia, early-morning awakening, or oversleeping
- Appetite and/or weight loss or overeating and weight gain
- Thoughts of death or suicide; suicide attempts
- Restlessness, irritability
- Persistent physical symptoms that do not respond to treatment, such as headaches, digestive disorders, and chronic pain

Mania

- Abnormal or excessive elation
- Unusual irritability
- Decreased need for sleep
- Grandiose notions
- Increased talking
- Racing thoughts
- Increased sexual desire
- Markedly increased energy
- Poor judgment
- Inappropriate social behavior

Causes of Depression

Some types of depression run in families, suggesting that a biological vulnerability can be inherited. This seems to be the case with bipolar disorder. Studies of families in which members of each generation develop bipolar disorder found that those with the illness have a somewhat different genetic makeup than those who do not get ill. However, the reverse is not true: not everybody with the genetic makeup that causes vulnerability to bipolar disorder will have the illness. Apparently additional factors, possibly stresses at home, work, or school, are involved in its onset.

In some families, major depression also seems to occur generation after generation. However, it can also occur in people who have no family history of depression. Whether inherited or not, major depressive disorder is often associated with changes in brain structures or brain function.

People who have low self-esteem, who consistently view themselves and the world with pessimism or who are readily overwhelmed by stress, are prone to depression. Whether this represents a psychological predisposition or an early form of the illness is not clear.

In recent years, researchers have shown that physical changes in the body can be accompanied by mental changes as well. Medical illnesses such as stroke, a heart attack, cancer, Parkinson's disease, and hormonal disorders can cause depressive illness, making the sick person apathetic and unwilling to care for his or her physical needs, thus prolonging the recovery period. Also, a serious loss, difficult relationship, financial problem, or any stressful (unwelcome or even desired) change in life patterns can trigger a depressive episode. Very often, a combination of genetic, psychological, and environmental factors is involved in the onset of a depressive disorder. Later episodes of illness typically are precipitated by only mild stresses, or none at all.

Depression in Women

Women experience depression about twice as often as men. Many hormonal factors may contribute to the increased rate of depression in women—particularly such factors as menstrual cycle changes, pregnancy, miscarriage, postpartum period, pre-menopause, and menopause. Many women also face additional stresses such as responsibilities both at work and at home, single parenthood, and caring for children and for aging parents.

A recent NIMH study showed that in the case of severe premenstrual syndrome (PMS), women with a preexisting vulnerability to PMS experienced relief from mood and physical symptoms when their sex hormones were suppressed. Shortly after the hormones were reintroduced, they again developed symptoms of PMS. Women without a history of PMS reported no effects of the hormonal manipulation.

Many women are also particularly vulnerable after the birth of a baby. The hormonal and physical changes, as well as the added responsibility of a new life, can be factors that lead to postpartum depression in some women. While transient "blues" are common in new mothers, a full-blown depressive episode is not a normal occurrence and requires active intervention. Treatment by a sympathetic physician and the family's emotional support for the new mother are prime considerations in aiding her to recover her physical and mental well-being and her ability to care for and enjoy the infant.

Depression in Men

Although men are less likely to suffer from depression than women, three to four million men in the United States are affected by the illness. Men are less likely to admit to depression, and doctors are less likely to suspect it. The rate of suicide in men is four times that of women, though more women attempt it. In fact, after age seventy, the rate of men's suicide rises, reaching a peak after age eighty-five.

Depression can also affect the physical health in men differently from women. A new study shows that, although depression is associated with an increased risk of coronary heart disease in both men and women, only men suffer a high death rate.

Men's depression is often masked by alcohol or drugs, or by the socially acceptable habit of working excessively long hours. Depression typically shows up in men not as feeling hopeless and helpless, but as being irritable, angry, and discouraged; hence, depression may be difficult to recognize as such in men. Even if a man realizes that he is depressed, he may be less willing than a woman to seek help. Encouragement and support from concerned family members can make a difference. In the workplace, employee assistance professionals or worksite mental health programs can be of assistance in helping men understand and accept depression as a real illness that needs treatment.

Depression in the Elderly

Some people have the mistaken idea that it is normal for the elderly to feel depressed. On the contrary, most older people feel satisfied with their lives. Sometimes, though, when depression develops, it may be dismissed as a normal part of aging. Depression in the elderly, undiagnosed and untreated, causes needless suffering for the family and for the individual who could otherwise live a fruitful life. When he or she does go to the doctor, the symptoms described are usually physical, for the older person is often reluctant to discuss feelings of hopelessness, sadness, loss of interest in normally pleasurable activities, or extremely prolonged grief after a loss.

Recognizing how depressive symptoms in older people are often missed, many healthcare professionals are learning to identify and treat the underlying depression. They recognize that some symptoms may be side effects of medication the older person is taking for a physical problem, or they may be caused by a co-occurring illness. If a diagnosis of depression is made, treatment with medication and/or psychotherapy will help the depressed person return to a happier, more fulfilling life. Recent research suggests that brief psychotherapy (talk therapies that help a person in day-to-day relationships or in learning to counter the distorted negative thinking that commonly accompanies depression) is effective in reducing symptoms in short-term depression in older persons who are medically ill. Psychotherapy is also useful in older patients who cannot or will not take medication. Efficacy studies show that late-life depression can be treated with psychotherapy.

Improved recognition and treatment of depression in late life will make those years more enjoyable and fulfilling for the depressed elderly person, the family, and caretakers.

Depression in Children

Only in the past two decades has depression in children been taken very seriously. The depressed child may pretend to be sick, refuse to go to school, cling to a parent, or worry that the parent may die. Older children may sulk, get into trouble at school, be negative, grouchy, and feel misunderstood. Because normal behaviors vary from one childhood stage to another, it can be difficult to tell whether a child is just going through a temporary "phase" or is suffering from depression. Sometimes the parents become worried about how the child's behavior has changed or a teacher mentions that "your child doesn't seem to be himself." In such a case, if a visit to the child's pediatrician rules out physical symptoms, the doctor will probably suggest that the child be evaluated, preferably by a psychiatrist who specializes in the treatment of children. If treatment is needed, the doctor may suggest that another therapist, usually a social worker or a psychologist, provide therapy while the psychiatrist will oversee medication if it is needed. Parents should not be afraid to ask questions: What are the therapist's qualifications? What kind of therapy will the child have? Will the family as a whole participate in therapy? Will my child's therapy include an antidepressant? If so, what might the side effects be?

The National Institute of Mental Health (NIMH) has identified the use of medications for depression in children as an important area for research. The NIMH-supported Research Units on Pediatric Psychopharmacology (RUPPs) form a network of seven research sites where clinical studies on the effects of medications for mental disorders can be conducted in children and adolescents. Among the medications being studied are antidepressants, some of which have been found to be effective in treating children with depression, if properly monitored by the child's physician.

Diagnostic Evaluation and Treatment

The first step to getting appropriate treatment for depression is a physical examination by a physician. Certain medications as well as some medical conditions such as a viral infection can cause the same symptoms as depression, and the physician should rule out these possibilities through examination, interview, and lab tests. If a physical cause for the depression is ruled out, a psychological evaluation should be done, by the physician or by referral to a psychiatrist or psychologist.

Treatment choice will depend on the outcome of the evaluation. There are a variety of antidepressant medications and psychotherapies that can be used to treat depressive disorders. Some people with milder forms may do well with psychotherapy alone. People with moderate to severe depression most often benefit from antidepressants. Most do best with combined treatment: medication to gain relatively quick symptom relief and psychotherapy to learn more effective ways to deal with life's problems, including depression. Depending on the patient's diagnosis and severity of symptoms, the therapist may prescribe medication and/or one of the several forms of psychotherapy that have proven effective for depression.

A good diagnostic evaluation will include a complete history of symptoms, that is, when they started, how long they have lasted, how severe they are, whether the patient had them before, and, if so, whether the symptoms were treated and what treatment was given. The doctor should ask about alcohol and drug use, and if the patient has thoughts about death or suicide. Further, a history should include questions about whether other family members have had a depressive illness and, if treated, what treatments they may have received and which were effective.

84.

Introduction to the Study of Experimental Medicine by Claude Bernard, and the Protection of Experimental Subjects

S tudies of the body's structure in health (anatomy) and disease (pathology) were the pillars of the healing arts and sciences for thousands of years. These, and related disciplines including microscopic anatomy and pathology (histology and histopathology), have been characterized by recent experts as "the dead hand of morphology"; the implications that while they are still essential for clinical medicine, they are not "cutting edge" and will provide no new breakthroughs. Aside from the cruel and thankless disrespect of such characterizations, there is a lack of appreciation for the development of medical science in which the focus changes from the structure to the ultrastructure to the molecular structure, and on and on to strings and beyond. No true science ever dies. Still, new science does develop, with increasing speed, and shoulders its way to center stage, shouting like a rude boy with a rachet in his waist, until a new science girl shows up with angel eyes that see forever. She will enchant us until tomorrow, when forever becomes yesterday's news.

One rude boy was experimental medicine. His rachet, the blunt instrument with which he made his way, was the laboratory. Recall Henry James's warning: "The effort really to see, and really to represent is no idle business in the face of the constant forces which make for muddlement" (see advance 42). There's nothing as full of those constant forces as the human body. By the middle of the nineteenth century, the need to isolate medical problems, and to reduce and control the number of forces confusing the issues, drove physicians into laboratories in increasing numbers. One of these was a discouraged French playwright turned physician-scientist. His name was Claude Bernard (1813–1878), and like many others, he based his work on animal experimentation. But laboratory and experimental medicine produced its own confounding forces.

Most of the experimentalists working with animals had little concern for their subjects' well-being. Conditions for laboratory animals were generally atrocious, giving rise to the antivivisection movement. In England, Queen Victoria's sympathy for the movement, along with a much publicized experiment on unanesthetized dogs, led to a cruelty prosecution and the 1876 Cruelty to Animals Act. In France, Bernard's wife, her dowery spent on his research, became an antivivisectionist and left him in 1870. French, British, and American experimentalists flocked to German labs where no such laws existed. Even where animal protection laws and organizations flourished, conditions remained miserable. And when, in the twentieth century, human experimentation began to grow, it was frequently associated with abuse and cruelty along racist or class lines, or excused by asserting that the subjects were going to die anyway. "Informed consent," where the subject is fully informed of the risks and benefits of the experiment and must give written consent, was not a viable concept until the late 1960s.

Nonetheless, the simplification and control provided by laboratory investigation allowed Bernard and others to make astonishingly rapid progress. Bernard's interests ranged far and wide among the physiological and pathophysiological workings of animals, organs, and cells. He experimented broadly, studying the role of poisons, such as curare's effect on neuromuscular transmission, which kills by respiratory paralysis and can be used medicinally, and carbon monoxide's ability to kill by displacing oxygen from hemoglobin. He studied the regulation of blood and body temperature, the neuronal control of vasodilation and constriction and thus on blood flow and pressure, and the role of the liver in maintaining blood glucose levels.

In 1865 Bernard published *Introduction to the Study of Experimental Medicine*. In his introduction he insisted that "scientific medicine, like the other science, can be established only by experimental means, i.e., by direct and rigorous application of reasoning to the facts furnished by observation and experiment."[114] Here he emphasized the importance of studying function directly by active intervention in the laboratory, of testing hypotheses in experimental animals under controlled conditions, of avoiding the constraint and passivity of hospital medicine as a venue for discovery, and of physiology over pathology as the way forward. "The ideas which we shall here set forth are certainly by no means new. . . . Our single aim is, and has always been, to help make the well-known principles of the experimental method pervade medical science."

And so Bernard did. Through his book, through the example of his work, and by the light of his meteoric career, scientific medicine was forever changed. His attention to experimental detail, as well as his grand conceptualizations, like the evolution of medical science and the maintenance of a stable internal environment (*milieu interieur*), advanced laboratory medicine, and they remain important influences to this day. Bernard was awarded chairs at both the Sorbonne and the French Museum of National History, became the president of the French Academy, and became a member of the senate. (Ah, only in France!)

The majority of great medical advances since Bernard's time have been made through scientific investigation. But in this setting as well, Apollo's gift of compassion has sometimes been ignored or worse.

Although the Nuremberg doctors' trials took place in December 1946, and the Nuremberg Code of Ethics, with its first principle, "The voluntary consent of the human subject is essential," came directly at the conclusion of the trials, the consent of human subjects was rarely obtained in the United States until the late 1960s. Even radiation experiments were being conducted by US government researchers for the precise purpose of observing what damage could be done.[115]

President Clinton's Advisory Committee on Human Radiation Experiments (ACHRE) was appointed in April 1995 to examine data relating to extensive US government experiments to observe any damage when Americans were injected or fed radioactive elements including plutonium, uranium, and radon. Pregnant women were given radioactive iron. None of these people had given their consent, most were never told of the experiments, and some had actually refused. These experiments, carried out many years after the promulgation of the Nuremberg Code, violated all ten principles designed to protect human subjects.[116]

In 1961 Harvard's medical school was faced with a dilemma. In order to obtain military research grants, it was asked to comply with the Nuremberg Code. But Harvard did not like the first principle. The administrative board at Harvard concluded that the Nuremberg Code is "not necessarily pertinent to or adequate for the conduct of medical research in the United States. . . . Faith and trust serve as the primary basis of the subject's consent. Moreover, being asked to sign a somewhat formal paper is likely to provoke inquiry in the subject, who can but wonder at the need for so much protocol."[117] This attitude typified the approach of the research community at that time.

This was to change in the late 1960s with the introduction of Institutional Review Boards for the Protection of Human Subjects (IRBs). It should be noted that I have been involved with IRBs and similar administrative structures since the time of their inception. I served on the Research Committee at the Bronx Veterans Administration Hospital in the early 1970s, the IRB at Montefiore Hospital, the Human Research Committee at Albert Einstein College of Medicine, and was the founding chair of the Animal Care and Use Committee (ACUC) at SUNY Downstate College of Medicine. In my experience, these committees work hard and with integrity. However, there are powerful competing interests that may affect the members and influence the environment. For the same valid reasons for which these committees were formed, we must strengthen local regulatory agencies.

Tissue Culture

A tissue culture is the successful growth of cells separate from the organism. The vital use of this concept—to study and manipulate individual cells, as opposed to an entire organism—is so ubiquitous in science today that the importance of the original discovery is often overlooked.

In 1907 embryologist Ross Harrison first grew a fragment of embryonic frog tissue in a drop of lymph fluid at Yale University. The drop hung from a cover slip, over a hollow slide, so that the tissue was enclosed in glass and could be watched under the microscope (later to be called the "hanging-drop" method). In so doing, he determined that nerve fibers grew from a single cell and not from a merging of many cells. He also proved what he called the "autonomous powers" of cells, which meant that when kept in aseptic conditions at the right temperature individual cells could live outside the body.

Taking Harrison's findings one step further, Alexis Carrel at the Rockefeller Institute for Medical Research introduced the idea of continuous culture. This is the idea that a living fragment of tissue could not only live for weeks but it could be grown eternally by using fragments of the culture to make new cultures ad infinitum.

86.

Understanding Alcohol and Health

I n considering alcohol-related illness, perhaps the greatest advance is the trend toward bringing our understanding of it into the open, removing the cloak of shame, and recognizing alcoholism as a treatable disease.

Alcohol, ingested as wine, beer, liquor, or pure grain alcohol, enters the stomach, and a small fraction is metabolized through the action of *gastric alcohol dehydrogenase*. Most then enters the small intestine where it is absorbed and travels in the blood plasma via the portal vein to the liver. There it enters the liver cells (hepatocytes) where it is metabolized through two pathways: hepatocyte alcohol dehydrogenase and the hepatocyte microsomal enzyme oxidizing system (MEOS). All of the alcohol is not metabolized in a single pass through the liver. Some passes through to exit via the hepatic vein and joins the general circulation. From there it distributes into the total body water and visits all the organs. Round it goes (recall William Harvey, advance 10), entering and reentering the liver where it continues to be metabolized to acid aldehyde and other products. It takes about forty-five minutes to metabolize half the alcohol in the plasma (metabolic plasma T-1/2). The peak concentration in the plasma determines the acute effects that the alcohol will have. This is primarily determined by the concentration and volume of the drink and the speed with which it is consumed. But other factors play a role as well. These include the speed of absorption, which can be influenced by other materials in the stomach (such as food), and the rate of hepatic metabolism. Hepatic metabolism increases in the chronic drinker as alcohol induces expansion of the MEOS, which is easily seen through a light microscope as swirls of microsomal tubules in the hepatocytes of a liver biopsy.

The acute effects of alcohol are mainly visited upon the brain as intoxication and the frequent consequence of rapid withdrawal known as a hang over, which can be modified by small amounts of alcohol, hence "hair of the dog that bit ya." (We're not suggesting you drink the next morning!) Acute alcoholic coma occurs when blood alcohol concentrations get extremely high. It is a very dangerous condition associated with high mortality and can be followed by withdrawal seizures. Acute alcoholic gastritis is a toxic inflammation of the mucosal lining of the stomach associated with retching and vomiting, sometimes with severe hemorrhage and hemetemasis (vomiting blood). Acute alcoholic hepatitis is a serious toxic inflammation of the liver, but its most ominous implication is its precursor relationship to alcoholic cirrhosis. These acute illnesses, if survived, should be taken as warnings of what is to come if drinking is not severely modified or ceased altogether. Pervasive television advertisements for whiskey and beer with those disingenuous little tags reminding us to drink "responsibly" don't help.

The chronic effects of alcoholism may involve the digestive, nervous, and endocrine systems. But the complexities of these involvements often result in impairment of blood clotting, damage to the cellular constituents of the blood (anemias and thrombocyteopenia), vascular distortions in the liver (portal hypertension), esophagus (esophageal varices), skin (spider angeomas), and elsewhere, all leading to chronic bleeding and sudden fatal hemorrhage. Jaundice, somnolence, and confusion (hepatic encephalopathy), and abdominal swelling (ascites) are among the host of symptoms that can result from alcohol-related organ dysfunction. These are commonly followed by death, and so understanding alcohol-related illness is a first requirement for preserving the lives of millions of people, including babies who can be born with fetal alcohol syndrome if their mothers drink during pregnancy.

The following information has proven helpful:

GENERAL FACTS

- Alcohol depresses the nervous system.
- Alcohol is the most frequently used brain depressant across all cultures and a significant cause of disease and death all over the world.
- Alcohol dependence runs in families and at least some of the transmission can be traced to genetic factors.
- The risk for alcohol dependence is three to four times higher in close relatives of people with alcohol dependence.
- Alcohol-related disorders are associated with a significant increase in the risk of accidents, violence, and suicide.
- It is estimated that approximately half of all highway fatalities involve a driver or a pedestrian who has been drinking.
- More than half of all murderers and their victims are believed to have been intoxicated with alcohol at the time of the murder.
- Repeated intake of high doses of alcohol can affect nearly every organ system.

TRENDS AND PREVALENCE

- The first episode of alcohol intoxication is likely to occur in the midteens. The onset of alcohol dependence peaks in the twenties to midthirties.
- Alcohol-related disorders typically develop by the late thirties.
- Alcohol-related disorders are among the most prevalent mental disorders in the general population.
- Approximately 14 percent of the population will suffer from an alcohol-related disorder at some point during their lifetime.

GENDER

- Alcoholism is more common among men than women across all cultures, with a male to female ratio as high as 5 to 1 (depending on age group and culture).
- Among women who drink alcohol, there has been a trend toward drinking heavily later in life.
- Once alcohol-related disorders develop in females, they tend to progress rapidly so that by middle age, females may have the same range of health, social, interpersonal, and occupational problems as males.
- Females tend to develop higher blood alcohol concentrations than males. As a result, women may be at greater risk for some of the health-related consequences of heavy alcohol intake.

ALCOHOL IN THE UNITED STATES

- As many as 90 percent of adults in the United States have had some experience with alcohol during their lifetime.
- Fifty-two percent of Americans over the age of twelve have drunk alcohol in the past month.
- Each year close to twenty thousand people die of alcohol-related causes in the United States. This does not include motor vehicle fatalities.
- Chronic liver disease is the twelfth leading cause of death in the United States.

SYMPTOMS OF ALCOHOL INTOXICATION

- Being overly talkative
- Impaired memory and judgment
- Inappropriate behavior, which may include aggression or sexual advances
- Loss of coordination and balance
- Mood swings
- Personality changes

- Slower reaction time
- Slurred speech

TREATMENT

A variety of treatments may be used:

- Alcoholics Anonymous or another program
- Individual therapy
- Family therapy
- Group therapy
- Vocational counseling
- Physician-prescribed medication (e.g., Antabuse, Naltrexone)
- Inpatient detoxification (if medical complications or the possibility of self-harm are present)

PRENATAL EFFECTS

Even small amounts of alcoholic beverages may be harmful during pregnancy:

- Consuming alcohol while pregnant may cause the mother to have a miscarriage.
- The baby may have a low birth weight.
- The baby may have mental retardation or developmental delays.
- The baby may be born with fetal alcohol syndrome, which is characterized by slowed growth, facial abnormalities, heart defects, joint and limb problems, and intellectual handicaps.

NEGATIVE CONSEQUENCES OF ALCOHOL ABUSE

- Brain damage
- Concentration, learning, and memory impairments
- Depression[118]
- Fifteen percent of heavy alcohol users suffer from liver cirrhosis and pancreatitis

- Hypertension (high blood pressure)
- Increased risk for cardiovascular disease
- Increased risk of accidents, violence, and suicide
- Increased risk of stomach cancer and cancer of other internal organs
- Korsakoff's syndrome (alcohol-induced persisting amnesiac disorder)
- Sexual impairment/dysfunction
- Suppression of the immune system

WARNING SIGNS THAT ALCOHOL IS A PROBLEM

- Continuing to drink despite significant problems related to drinking (e.g., car accident)
- Craving alcohol
- Downplaying one's alcohol consumption, including frequency and intensity
- Drinking alone
- Needing to consume more alcohol to gain the previously experienced high
- Reducing or stopping important social, occupational, or recreational activities because of alcohol use
- Spending a great deal of time in activities necessary to obtain alcohol or to recover from its effects
- Unsuccessful attempts to control or cut down alcohol intake

87.

Preprofessional Healthcare and the Development of Fee-for-Service

Restore a man to his health and his purse lies open to thee.

—Robert Burton (1577–1640), English clergyman and scholar

Over the course of human history, including in most of the developed world today, societies have shown that this vulnerability does not have to be exploited. Profiting, or making a living, from the misfortunes of others may be unavoidable, but profiteering should be easily defined, unacceptable, and preventable.

Healthcare in the twenty-first century, outside of the United States, is increasingly seen as a basic human right, one that deserves to be protected and provided at an affordable fee to all citizens of civilized societies. This idea—that medical procedures and healthcare in general should not be subject to or motivated by market forces—is one that has, in recent decades, evolved back into favor only after repeated experiments with the capitalization of healthcare led to systematic and catastrophic failures, resulting in grotesque profits on the supply side contrasted with the suffering of millions of disenfranchised patients on the demand side of the equation.

The first known healers were the shamans, also known as medicine men (although they were usually women), whose services were valued in cultures across the globe dating back over forty thousand years. According to the *Cambridge Encyclopedia*, a shaman is "[a] person to whom special powers are attributed for communicating with the spirits. . . . The spirits help him do his chores which include discovering the cause of sickness, hunger and any disgrace, and prescribing an appropriate cure." Shamans did not accept payment for their services in either money or gifts, because it was believed that payment was already rendered by the spirits.

However, as bartering gave way to a moneyed system throughout developing societies, healthcare was no exception. The earliest recorded history of medicine as practiced today begins with the Egyptians (c. 2500 BCE) and the Code of Hammurabi. Inscribed into a huge stone pillar, this eclectic grouping of laws defined everything from the conduct of physicians to punishments for malpractice. It also set precise rules regarding payment for medical services rendered. Law 215, for instance, states, "If a physician makes a large incision with an operating knife and cure it, or if he opens a tumor [over the eye] with an operating knife, and saves the eye, he shall receive ten shekels in money." Law 221: "If a physician heals the broken bone or diseased soft part of a man, the patient shall pay the physician five shekels in money." At approximately the same time, the Hebrew civilization codified similar practices in the Pentateuch and Talmud as prescribed by Moses.

It wasn't until the late Middle Ages (first to fifth centuries CE) that Western society saw the gradual professionalization of medicine. Medical guilds, established across Europe, determined for the first time exactly which people could perform which tasks: barbers could cut and bleed; executioners could set bones; and women, among the first healers in human history, could do little more than deliver babies. Under the protection of a guild, physicians practiced medicine with less fear of the repercussions from the outside world. Additionally, in many cases, guilds would encourage treatment of all patients, the poor and the rich, the same. The charge for service was according to what the patient could pay.

In the fourteenth century, professionalization continued to grow as French and German towns began to license their doctors. But the movement was progressing slowly—so slowly that by the end of the fourteenth century, Paris still had only ten licensed doctors for more than two hundred thousand people.

Over the next several centuries the business of healthcare expanded along three divergent roads: guilded providers, charity-based providers, and fee-for-service doctors. The relative dominance of each varied across Europe, but, in general, fee-for-service—where patients could select the doctor of their choice at a price determined by market forces—became the standard. This freedom of choice, however, had many predictable consequences, with inequality of medical services between economic classes being the most glaring. Nonetheless, this inequality had a relatively small effect until the nineteenth century when advances in diagnosis and treatment began to have an increasing impact on the understanding and outcomes of illness and injury. At that time, demands for equitable access to medical care intensified. In Great Britan the foundations of the National Health Service were laid by works such as Kay-Shuttlesworth's *The Moral and Physical Condition of the Working Classes Employed in the Cotton Manufacture* (1832) and Chadwick's *Report on the Sanitary Condition of the Labouring Population of Great Britain* (1842), which pointed out that that the average age of death among laborers in London was sixteen, while among the higher classes it was forty-five—conditions that would be vividly depicted by the great Charles Dickens.

Government entered healthcare with the first British Public Health Act in 1848, which created a central authority, the General Board of Health. During World War II the Emergency Medical Service took over most of Britain's hospitals, and the improvement in efficiency and service mandated a continuation of government control. In 1948 the National Health Service, Britain's socialized system, was inaugurated.

At the end of the nineteenth century, American companies pioneered the contracting of medical services. This innovation came from industries such as railroad and lumber, whose operations often took place in isolated areas and whose employees would otherwise have no access to medical care. Some companies went so far as to build and operate their own hospitals, and they staffed those hospitals with physicians who were paid out of the company payroll. But the growth of fee-for-service medicine marched on. In 1847 the formation of the American Medical Association (AMA) prevented rational discussion of any arrangement other than fee-for-service. The AMA supported segregation, restriction of women in medicine, and advertising. The AMA almost bankrupted itself fighting Medicare and Medicaid, calling these and other social programs that broadened access to healthcare "communism" and branding anyone who would dare even discuss them a communist.[119]

Life expectancy and infant mortality rates—two of the best indicators of overall health—tell a story that suggests that alternatives to the United States' current system must be considered. Average life expectancy in Great Britain was 77.4 years in 1998. Life expectancy for the US population reached 76.9 years in 2000. Infant mortality in Finland is below 4 percent; in the United States it is 7 percent. Health services are available to all in Finland, regardless of financial situation. In fact, the United States is nowhere near the top in vital statistics related to health, and it spends far more on healthcare than most other countries.[102] Consider the following: Americans spend considerably more money on healthcare services than any other industrialized nation, but the increased expenditure does not buy more care, according to a study by researchers at the Johns Hopkins Bloomberg School of Public Health.

The study found that the United States spent 44 percent more on healthcare than Switzerland, the nation with the next highest per capita healthcare costs, in the year 2000. At the same time, Americans had fewer physician visits, and hospital stays were shorter compared with most other industrialized nations. The study suggests that the difference in spending is caused mostly by higher prices for healthcare goods and services in the United States. The results are published in the May/June 2003 edition of *Health Affairs*.

"As a country, we need to ask whether increased spending means more resources for patients or simply higher incomes for health care providers," said Gerard Anderson, lead study author and professor in the school's departments of Health Policy and Management and

International Health. "Policymakers should assess exactly what Americans are getting for their greater health care spending. In economics, these are known as opportunity costs because you can spend the money in different ways," said Dr. Anderson.

For the study, Dr. Anderson and his colleagues compared health systems data of the thirty industrialized countries in the Organization for Economic Cooperation and Development (OECD) from the year 2000, which are the most recent data available. The authors examined the factors contributing to higher healthcare prices in the United States. They also compared pharmaceutical spending, health system capacity, and use of medical services.

According to the study, US per capita health spending rose to $4,631 in 2000, which was an increase of 6.3 percent over the previous year. The US level was 83 percent higher than Canada and 134 percent higher than the median of $1,983 in the other OECD member nations.

The researchers found that the healthcare spending gap between the United States and other industrialized countries widened between 1990 and 2000, despite efforts to control spending with managed care.

The study also found that the United States spent 13 percent of the nation's gross domestic product (GNP) on healthcare in 2000, which was considerably higher than other nations. In contrast, Switzerland spent 10.7 percent of its GNP on healthcare, while Canada spent 9.1 percent. The median spending level for the OECD nations was 8 percent. American private spending per capita on healthcare was $2,580, which was more than five times the OECD median of $451. In addition, the United States financed 56 percent of its healthcare from private sources, which was the greatest amount of the OECD countries.

According to the study, public financing of healthcare from sources like Medicare and Medicaid accounted for 5.8 percent of the US GDP in 2000, which is similar to the OECD median of 5.9 percent. However, the United States spent $2,051 of public funds per person, which was much more than the OECD median of $1,502. In most of the other OECD countries the public healthcare expenditures cover everyone, unlike the United States.

Socialized Medicine

Real socialism is inside man. It wasn't born with Marx.
—Dario Fo (b. 1926), Italian playwright

The socialism of our day has done good service in setting men to thinking how certain civilizing benefits, now only enjoyed by the opulent, can be enjoyed by all.
—Ralph Waldo Emerson (1803–1882)

For a concept of such simplicity, the idea of socialized medicine is misunderstood with surprising regularity in the United States. In an effort to put to rest the two biggest misconceptions (i.e., socialized medicine is akin to a single-payer system; socialized medicine demands a socialized economy), we present the *American Heritage Dictionary*'s quick and complete definition of the term: "A system for providing medical and hospital care for all at a nominal cost by means of government regulation."

Because the definition lacks any ambiguity, it is interesting to investigate the cause of such widespread confusion. Over the past fifty years, the weighted term *socialized medicine* has been wielded frequently as a rhetorical noose, cleverly wrapped around the neck of any egalitarian system that would serve to marginalize profits. Critics of the concept invariably focus on the inherent lack of choice associated with government-provided health services. Because every aspect of a socialized healthcare industry is controlled by and provided by the government—all doctors, nurses, medics, and administrators are government employees—the system, such as the National Health Service in England, determines where, when, and how services are provided.

Upon further investigation, however, this patient-service relationship does not necessarily correspond to inferior care or decreased satisfaction. According to a report by the US-based Health Affairs, *Common Concerns amid Diverse Systems: Health Care Experiences in Five Countries*, the United Kingdom's socialized medical system outperforms the US system in patient-reported perceptions. In other words, the people with direct experiences report greater satisfaction with their health services under a socialized system than they do in a free-market system. These results appear all the more surprising if we consider the fact that the US per capita healthcare expenditures ($4,887) are nearly two and a half times those in the United Kingdom ($1,992).

The British system is probably the most instructive example for Americans to evaluate because of the similarities in economy and government structure between the two nations. According to www.nhs.uk.com, the National Health Service "was set up on the 5th July 1948 to provide healthcare for all citizens, based on need, not the ability to pay." Originally conceived as a response to the massive casualties of World War II, the system survives and continues to evolve to this day. The NHS is funded by taxpayers and managed by the Department of Health, which sets overall policy on health issues. Individual patients are assigned a primary care center (which provides doctors, dentists, opticians, pharmacists, and walk-in centers) managed by a primary care trust (PCT). The NHS explains its system of referrals this way: "If a health problem cannot be sorted out through primary care, or there is an emergency, the next stop is hospital. If you need hospital treatment, a general practitioner will normally arrange it for you."

PCTs are responsible for planning secondary care. They look at the health needs of the local community and develop plans to set priorities locally. They then decide which secondary care services to commission to meet people's needs and work closely with the providers of the secondary care services who deliver those services. It is important to note that the definition of socialized medicine allows the system to adapt to any and all of the unique requirements of a particular country's patients.

The NHS may be the world's most sophisticated socialized medical system, but the modern world's first was established by the former Soviet Union in the 1920s. China, Cuba, Sweden, and most of Scandinavia have successful and completely socialized healthcare systems today.

This presents a puzzling question: If this system is cheaper and at least as effective, why has socialized medicine been so successfully and systematically vilified in the United States? Perhaps it is in name only. Maybe this idea would be easier for Americans to swallow if it was stripped of its anticapitalist prefix. Perhaps simply swapping "socialized" with "public" would do the trick. It certainly has worked well for other government-funded programs in the United States, such as welfare, libraries, schools, parks, and interstate highways.

89.

Single-Payer Systems

L ike socialized medicine, the definition of a single-payer healthcare system is simultaneously unambiguous and almost universally misunderstood. Quite simply, "single-payer healthcare" describes a system in which healthcare costs are financed through a single source for a nation's entire population. The capital in such an arrangement is most often collected through a progressive tax on individuals and businesses.

What single-payer is not, as is frequently insinuated by those financially invested in the current US system, is socialized medicine (a system in which the government owns, operates, and provides every aspect of the healthcare services). While it is true that in a single-payer system the government collects and disperses the capital for services rendered, its decision-making responsibilities end there. Contrary to another deliberate misconception, a single-payer system *does* allow individual citizens to choose their physician and demand all appropriate medical tests and procedures without out-of-pocket expenses.

Physicians in a single-payer system, as they are in our current system, are financially compensated for their services on a fee-for-service basis, billing the trust according to previously established fee criteria. Capital improvements, such as the construction of new hospitals, are provided through a separate fund.

Where does the money come from? Tellingly, the answer to that question is often debated, despite the reputation of its source. Studies by the US Congressional Budget Office and the General Accounting Office show that single-payer universal healthcare would save $100 billion to $200 billion per year, while covering every currently uninsured American *and* increasing healthcare benefits to those already insured.[121]

How is this possible? Unfortunately for those in the health insurance industry, a large portion of those savings come at the expense of their profits. According to a 2000 American Medical Student's Association study, the overhead and profits of the nation's seven biggest HMOs accounted for more than 21 percent of their premiums. Over the same period of time, Canada spent just 1 percent of its citizens' money on overhead, and the only profits were made by physicians. In America the net income for the nation's health insurers was $10.2 billion in 2003, according to Weiss Ratings, Inc.[122] That's nearly double the profits earned in 2002 and almost fourteen times the $736 million earned in 1999. Over that time, net profit margins for the health insurance industry rose to 3.78 percent from 0.38 percent, a tenfold increase. The US General Accounting Office concluded that the streamlining which would occur in a single-payer system would immediately generate $34 billion in savings from insurance overhead alone. It went further to conclude that if "the universal coverage and single-payer features of the Canadian system were applied to the United States, the savings in administrative costs alone would be more than enough to finance insurance coverage for the millions of Americans who are currently uninsured. There would be enough left over to permit a reduction, or possibly even the elimination of co-payments and deductibles."

Further highlighting this point, a recent study by Harvard Medical School researchers found that, even without taking into account all the money diverted toward corporate profit, for-profit hospitals spend 34 percent of their total budget on administration, compared to

24.5 percent for nonprofit hospitals.[123] But disregarding the money diverted toward corporate profits would be a mistake; CEO Leonard Abramson alone received $967 million the day US Healthcare merged with Aetna in 1996.[124]

At the same time as Americans are paying huge health insurance premiums (the United States spends at least 40 percent more per capita on healthcare than any industrialized nation with universal healthcare), the level of care remains inadequate and worsening, as comparative statistics clearly show. The United States ranks twenty-third in infant mortality worldwide, down from twelfth in 1960. It ranks twentieth and twenty-first in life expectancy for women and men, respectively, down from first in 1945. And the US immunization program ranks sixty-seventh, right behind Botswana.[125]

The most puzzling aspect of the US government's refusal to convert to a single-payer system is that it already operates a successful single-payer healthcare system: Medicare.

Without qualification, Medicare, which provides universal coverage for the nation's forty million elderly and disabled, is a classic single-payer system, with overhead costs at just 3 percent of premiums as opposed to the previously stated 21 percent for HMOs.

90.

The Fight to Open Medicine to Women and Minorities

There have been six doctors in our family. All men. We entered medical school before "the great leap forward" in which women began applying. Before 1970 women were fewer than 10 percent of medical school applicants. Now they represent about 42 percent. And they get accepted at a rate of about 47 percent, at virtually the same rate as men. In 1970 women made up about 8 percent of all US physicians. By 1990 it was 17 percent, and it will be 30 percent by 2010. That's progress.

Among underrepresented minorities (blacks, Hispanics, American Indians, Asian Pacific Americans), African Americans represented, in 1990, 11.8 percent of the US population, and after decades of effort to improve their representation in medicine, they made up only 3.6 percent of US physicians. This underrepresentation of minorities in medical professions and access to care is not improving, and it is manifested in shocking statistics defining minority health status.[126] Infant and maternal mortalities, life expectancies, death rates from prevent-

Fig. 51. Many witches who were burned in medieval Europe were women healers.

able and curable diseases, and so on all define an undeniable reality that must be addressed.[127]

If you accept what we have affirmed about Apollo's greatest gifts, the idea of healing (advance 1), and the doctor-patient relationship (advance 2), then it follows that the healing arts and sciences must be open to qualified candidates of all backgrounds, races, religons, and both sexes. That this should be clear from many timeworn principles of ethics, religion, science, and morals does not alter the fact that medicine has been closed to women and minorities. On the contrary, it should cause us to examine the origins and causes of this backwardness and to appreciate the profound advance that is represented by redressing this centuries-old affront to Apollo. This restricted access hurts you and everyone who wants decent healthcare. Not only does it narrow the availability of physicians, but it restricts the focus of research and attention to the full range of healthcare problems and solutions.

Women, and people classified in repressive societies as "minorities," have always practiced healing. Wisewomen gathered herbs, brewed them into medicinal potions, and gave the nursing care that was *the* major help available for the sick until modern times. Most of the remedies provided by physicians were ineffective until about a hundred years ago. But women always bathed the arthritic and manipulated their joints, and looked after pregnant women and delivered babies. For most of human history healing was the concern of women.

In ancient Egypt women were prominent physicians, notably in the medical schools at Heliopolis and Sais. "Agamede of the golden hair" in Homer's *Iliad* was skilled in medicine and herbal lore. In classical Greece, Philista (318–372 BCE) lectured so effectively that pupils flocked to her. Eudoxia—"empress and wife of Emperor Theodosius"—founded a great hospital in Jerusalem in the fifth century CE. Empress Julia Anicia (472–512) was deeply interested in medicine; her illustrated herbal volume has survived to this day. Women physicians were numerous during Roman times, though few of their writings have been preserved. But there are a significant number of tombstones that read something like "To Primilla, my sainted goddess, a medical woman."[128] In the same period women physicians were common among the Germanic "barbarians."

During the Middle Ages, the learned of both genders in Christian monastic institutions preserved herbal and diagnostic skills. The best-known woman among them was the mystic Hildegarde of Bingen (1098–1179). She was the tenth and last child of a noble family that offered her to God as a tithe. She had visions, explained to her in Latin, she said, by a voice from heaven. She wrote two medical manuscripts: on plant, animal, and mineral medicines; and on physiology and the nature of disease. Her remedies were partly herbal and partly spiritual or magical.

From the seventh century on, new hospitals opened that were controlled by the church and organized by nuns. This was a formal beginning to the clear division of healing duties along gender lines. Women healers were clearly a threat to European male dominance. Aside from benefiting from the emerging fee-for-service system in which women represented an income drain from officially sanctioned male physicians, women were generally empiricists who learned from experience rather than by accepting the "revealed wisdom" passed on through the church. The guilds of male physicians and surgeons, which were forming at the time, did not admit women applicants. In 1322 Jacoba Felicie appeared for examination before the all-male Paris guild, and although she was allowed to practice, she was among very few women to be licensed at the time. The Catholic Church and its Inquisition, beginning in the fifteenth century, persecuted and murdered witches by the thousands, many of whom were skilled herbalists and healers. *The Witch Hammer* (The Malleus Malefi-carum)[129] published in 1486 was the twisted bible of this misogynistic church movement that led to the deaths of so many innocent people. The last witch was burned in Germany in 1775, but *The Witch Hammer*, with its murderous defamation of women, was hailed by the church well into the twentieth century.

In eighteenth-century Europe, the medical profession was developing a more formal and rigid structure. Guilds of physicians were largely male and did not admit women to their apprenticeships. Over the centuries they built up and guarded their exclusivity. The European system was enthusiastically followed here in America. In 1858 the new *English Medical Register* contained the name of one woman. When Elizabeth Blackwell graduated in 1849 from Geneva Medical College of New York, she became the first woman medical school graduate of modern times. It was not coincidental that her father was an abolitionist. A few years later the Women's Medical College of Philadelphia was founded. Scant progress was being made in the United States for many years. The Popular Health Movement of the 1830s and 1840s,[130] with its efforts to open the profession, was beaten down with the whip of the American Medical Association formed in 1848. The AMA was composed of only white male professionals. In 1910 the Carnegie Corporation and the Rockefeller Foundation produced the famous *Flexner Report*, "the foundation of American medical education," which resulted in the closing of most of America's black medical schools and the majority of those that admitted women. The Carnegie Foundation in conjunction with the AMA Council on Medical Education hired former schoolmaster Abraham Flexner to report on "the proper and actual basis of American and Canadian medical instruction for immediate action." The Carnegie Foundation published Flexner's accounts, *Medical Education in the United States and Canada: A Report to the Carnegie Foundation for the Advancement of Training*,[131] popularly called the *Flexner Report*. In it, Flexner summarized in a direct and critical approach the 168 medical schools he visited. He called for the closure of all three medical schools that had all women students. He subsequently advocated the reduction of the seven African American medical schools to two, for he felt that "medical care to the [African American] race should never be wholly left to [African American] physicians." All these "reforms" came from on high, and the citizenry had little knowledge or understanding of what transpired.

There is gender and racial bias built directly into the structure of medicine as it currently exists. In both the academic and the clinical spheres, medicine is structured as a hierarchy with powerful white male figures controlling opportunities for training, practice, recognition, promotion, and virtually all aspects of development. The structure is based on the male life cycle, to a linear career progression with no concern for childbearing or nurturing—male standards of behavior predominate. This atmosphere, particularly in the early and most impressionable phases of careers, is rather like an old-fashioned ball team or combat unit with small groups working long hours under stressful conditions. When social barriers are strained, women and minorities are expected to adhere to the rules and go along as "team players." We have seen dominance, ambition, and competitiveness prized. Sensitivity, responsiveness, deference, and accommodation are viewed more as characteristics of a social worker. In 1992 the Association of American Medical Colleges asked graduates if they had been subjected to sexual

Fig. 52. Dr. Donald Wilson.

harassment or gender discrimination during medical school. Sixty percent of women said yes. A study of five successive classes at a midwestern medical school revealed that 34 percent of the women had personally experienced gender discrimination and 62 percent had observed discrimination against classmates. And now that giant corporations control "the industry" and pay doctors modest salaries to see thirty to forty patients per day, women are getting more than their share. Women physicians earn about 40 percent less than their male counterparts.

There are 126 medical schools in the United States. The number of female or minority department chairs or deans is exceedingly small, and in each category they can be counted on the fingers of one hand. The exception: there are ten women chairs of pediatrics. Ninety-three percent of tenured full professors in clinical departments of American medical schools are men.

Recently the media trumpeted Harvard University president Lawrence Summers's alleged statement that genetic differences may help explain why fewer women than men succeed in science and math careers. In apologies, Summers said his talk was misconstrued and that he does not "believe that girls are intellectually less able than boys, or that women lack the ability to succeed at the highest levels of science."[132] Geraldine Richmond, who holds an endowed professorship in chemistry at the University of Oregon, heard of Summers's purported remarks from colleagues who heard him speak. She called his comments "remarkably stupid."

Many people still struggle to advance these essential causes. One of today's most prominent leaders is Dr. Donald Wilson, dean of the University of Maryland School of Medicine.

91.

Health Maintenance Organizations

As we mentioned in our introduction, not all the "greatest" medical advances are regarded as great in the positive sense of the word. Some achieve greatness by mere dominance. Did the Persians, for instance, regard the king of Macedon as great? Probably not. Yet Alexander was able, through brutal successes, to forever alter the human landscape, and greatness was bestowed on him. This is how we see the invention of HMOs.

No one would agrue against the fact that health maintenance organizations (HMOs) now dominate the healthcare marketplace. HMOs are a legal entity that assumes responsibility for the healthcare services of an individual or group in return for a prepaid monthly premium. Only those medical professionals designated by the HMO may be contracted for service, and all visits, prescriptions, referrals, and other care must be cleared in advance by the HMO. The participating medical professionals are organized according to several similar models: staff, group, hybrid, network, and POS (point of service). In both the staff and the group models, all approved medical professionals are salaried exclusively by the HMO, while a network model allows approved medical professionals to work in their own offices while not working exclusively for the HMO.

A preferred provider organization (PPO) is a healthcare delivery system in which the employer or insurer enters into contracts with a limited number of healthcare providers (physicians, hospitals, etc.) in order to provide healthcare services at a discount. The preferred providers are often subject to specific instructions regarding "monitoring of utilization, the appropriateness of care provided, and the terms of the provision of care allowed under the arrangements." Patients may have the choice to use nonpreferred medical providers but there is always a financial incentive to use the PPO.

The fact is that in the decade since HMOs have achieved that domination, healthcare costs have skyrocketed (see advance 87). The question is: What value do they add to the product?

92.

Preventive Medicine

"An ounce of prevention is worth a pound of cure."

—Benjamin Franklin

Nowhere is the old adage above more apt than in the field of health. Preventive medicine is the approach to illness that emphasizes preserving health, and we celebrate it for its undoubted, if often undocumented, contributions to human well-being. People have always practiced prevention as best they could, whether it was a mother dressing her child warmly to help him avoid catching a cold or a "plague doctor" dressing in costume to ward off the plague.

Plague doctors notwithstanding, the medical profession has been severely criticized for being insufficiently interested in preventive medicine. To this I must plead, on my own behalf, and on that of my colleagues, "guilty with an explanation." Set aside for a moment the charge that the profession has an inherent conflict of interest with respect to prevention. I think that this is overly cynical, and I have never seen evidence of this, certainly not among the thousands of physicians with whom I have had contact.

But the accusation is not to be dismissed out of hand, if only because of the historical actions, and inactions, of our guild, the AMA, in support of the tobacco industry (see advance 33).[133] Nevertheless, preventive measures require understanding of the cause and progres-

Fig. 53. "Plague doctors" in costumes that were worn to prevent physicians from catching the plague.

sion of disease, as well as methods of intervention. Great strides have been made in these areas (many are discussed in this book), but actual prevention also requires commitment on the part of individuals and governments, as well as healthcare professionals. The establishment and maintenance of public works (advance 21), including water supplies (advance 21), sewage systems (advance 20), and garbage disposal systems (advance 93), are essential to the prevention of widespread disease. Drainage of standing water and other mosquito-control measures are also governmental responsibilities capable of preventing epidemic disease. And governments have done heroic work in these areas, although when they fail us, as in the case of New York City's mosquito-control program in the 1990s, they court disaster. But governments have taken on and succeeded with monumental tasks such as lowering the levels of rivers and streams to kill mollusks that live at the water line and transmit schistosomal diseases. And the WHO's success in eradicating smallpox is to be celebrated (advance 23).

Some preventive measures are cheap and easy, such as giving out bed netting to prevent malaria, but they don't always get done. Others, like cessation of smoking, and diet and obesity reform, are more complex, with necessary roles for industry, government, and the public in order to make progress.

The point here is that many of the big killers, like AIDS, malaria, infectious diarrheas, type 2 diabetes, coronary artery disease, stroke, lung cancer, colon cancer, and breast cancer, are largely preventable. If there's the will, there's the way.

93.

Refuse Disposal

Garbage is dangerous, and people die every day through ignorance regarding the destructive powers of garbage. Throughout history millions of people have died because their garbage was not properly disposed of, and it was allowed to fester. The great plagues are legendary. There was the one that killed 40 percent of Constantinople in the sixth century CE, and the infamous Black Death in fourteenth-century Europe killed twenty million people in just four years. And, despite technology that could end this phenomenon, close to two thousand people died from plague in India during the early 1990s and more than ten million died in the twentieth century.

Every one of these deaths could have been avoided by the proper disposal of garbage. Basic sanitation practices would reduce rat populations that harbor the lethal bacteria, pesticides can kill both the rats and the fleas that spread the disease, and there are potent antibiotics to treat the victims. But ignorance is a deadly disease.

In the United States such dangers do not exist. We collect 3.9 million tons of municipal solid waste every year, most of it finding its way to the nation's 2,500 landfills.[134] Unfortunately, even when garbage is

collected properly and deposited in landfills, there are serious potential health disasters. The New York State Department of Health
found that women living near solid waste landfills where gas escapes
have a fourfold increased chance of bladder cancer or leukemia.[135]
And there isn't a landfill on Earth that isn't leaking gas. Landfill gas
consists of naturally occurring methane and carbon dioxide, which
form inside the landfill as the garbage decomposes. As the gases form,
pressure builds up inside the landfill, forcing the gases to move. Some
of them escape through the surrounding soil into the water supply
and others simply move upward into the atmosphere. In addition, a
survey conducted by the EPA in the late 1980s found that half of all
municipalities will run out of landfill space within ten years and that a
third of all municipalities will run out within five (the EPA was correct on both predictions).

The most effective alternative to landfills is incineration plants.
These plants may emit harmful fumes as well, but nothing compared
to the disastrous long-term effects of overflowing landfills. The
problem then becomes what to do with the incinerated ash that contains hazardous levels of arsenic, lead, and mercury. In fact, more than
fifteen thousand tons of such ash from municipal incinerators in
Philadelphia was stuck on a moving barge for more than sixteen years,
rejected by eleven countries, five states, and the Cherokee Nation.[136]

The solution will most likely be a combination of methods. Big
garbage incinerators, which also generate steam for electricity, are
capable of housing recycling programs. Environmentalists are also
considering outfitting landfills with impermeable liners to prevent the
pollution of groundwater in addition to monitors for air and water
pollution. Such landfills are used to dispose of toxic waste but thus far
have not been used for garbage. Each of these alternatives has its own
economic or environmental problems. Still, incinerators that can
reduce the volume of trash by up to 90 percent while producing
energy to be sold to public utilities are certainly a viable option. In the
late 1980s there were only 100 such incinerators in use in the United
States, consuming about fifty thousand tons of solid waste a day. That
number jumped to 150 by 1993 but fell to just 97 by 2001, mostly
because of the high cost of operation.[137]

94.

Recalled to Life by Physical and Occupational Therapy

Once you fully apprehend the vacuity of a life without struggle, you are equipped with the basic means of salvation.

—Tennessee Williams

A few years ago, when I was sixty years old, I suffered a stroke that resulted in paralysis of my right side. While trying to recover, I was bitten by a tiny venomous spider called the Brown Recluse (*Loxosceles reclusa*), which caused necrosis (death) of the tissues of my left hand and arm. In an attempt to save that arm, I endured ten operations and had it slung up in a vertical position. The immobilization complicated the stroke with further weakness and frozen joints, so that after about two years in hospitals, I could move my head only five degrees in each direction—no walking, no feeding myself, no changing position, no scratching itches. I did a lot of thinking.

Now, I'm not recommending this, but for a physician there can be nothing more instructive than a near-fatal illness that brings him to medical and surgical wards, intensive care units, radiology departments, waiting in halls on a gurney, and lying in bed at 4 AM having to urinate but unable to ring for the nurse. I had studied hospital medi-

Fig. 54. Physical therapist performing passive stretching exercises on a patient's right leg.

cine from the ground up, and, in a similar way, I now learned about physical medicine and rehabilitation by doing. Faced with my case, therapists often smiled, gave me a warm welcome, and said, "Well, let's see, where do we begin?"

There have been basic scientific advances and technologic innovations that have helped to bring the disabled from being stored in warehousing facilities to living a productive life. The most important of these derive from the appreciation, gained through many observations and scientific studies, that the brain is a dynamic, adapting structure, capable of restoring significant lost function. A generation ago we were taught that the brain was rather like a switchboard: if you damaged the wires, or neuronal tracts and nerves, that was it—you lost function, never to return. Now we can scan the brain and see new pathways developing under the stimulus of intension and struggle.

Physiatrists are physicians who specialize in rehabilitation medicine. They know the science, do research, teach, and direct the physical medicine and rehabilitation services. Therapists know the patients best, and they do most of the motivation and the daily work that produces results. Physical therapists are primarily concerned with walking, speech therapists with talking, and occupational therapists with the function of arms and hands, although there is considerable crossing of these lines.

Because my stroke was complicated by extensive retroperitoneal bleeding and a severe attack of gout, I did not see a therapist for many weeks. Lying in bed unable to move my right side at all (except when I yawned, whereupon the arm would rise six inches off the bed, causing uncontrollable glee, and more yawning), I began to appreciate my intact functions: consciousness, memory, intellect, and speech. At least I wasn't "locked in" like the hero in Dalton Trumbo's World War II novel, *Johnny Got His Gun*. But walking and using my hands were becoming unimaginable. To stave off the inevitable depression, which I figured was coming around the corner with the morning milk, I busied myself with family and friends. Their companionship was uplifting. Their love was lifesaving.

And then one very fine day two physical therapists came to my bedside and asked, "Are you ready to stand up?" I thought they were quite mad. Had they not looked at my chart and seen that my body was useless? But I could hardly say no, since I was well known there and didn't want to be seen as a bad or cowardly patient. "Sure thing," I said, smiling, "I'll just pop up."

Before I could begin valiantly struggling in vain, they laughed and insisted, "Not so fast. You'll need help. First, you will just sit with your feet hanging over the side of the bed." And with great strength and care these two women slowly moved me into a sitting position, warning that I might become dizzy, but they would support me so I would not fall.

I sat at the side of my bed, my body waving in an imaginary wind like an infant, like a wet noodle, a frightened chicken too bewildered to cluck.

"Great," they cried. "Excellent, very good, see how long you can hold it there, and if you need to lie back down, just give the word. You're in control."

"I don't feel in control," I said.

"But you are," they said. "Try to sway from side to side."

"But I am," I said.

"Try to control it, exaggerate the movement slightly, back and forth, we're right here," said Laura, the younger therapist, who was about the age of my daughters. So I tried hard, but I slumped onto Violet, who propped me up saying, "Wonderful, good. Now lean on me and rest."

"Are you ready to stand up?" Violet asked.

I was going to ask if she was serious, but by then I knew that they were very serious people, and they were already standing and holding me front and back, left and right, and I felt my left leg trying to lift my dead weight. Miraculously, I was moving up, but also forward and back a bit, enough to activate the freight reflex, and slumping to the right, shooting adrenalin around the room until I was upright-ish.

"Whoa, hang on!" I cried.

"Try to hold it right there," they urged.

I tried, oh how I tried, for about two seconds. Then I sank down as they guided me back to a sitting position amid their cheers. Had I been nominated for a Nobel Prize? It felt good, but I had had it.

"One more time," said Laura, with a wink. They are nuts, I thought, or sadists. But they were doing what I figured was almost all the work. How could I say no? So, up I went, this time holding it for three seconds before getting ready to plop when Laura demanded, "Let's take a step!"

Somehow I put my left foot forward and dragged my right side along. It must have been ugly. It was not independent. It certainly wasn't walking. But I stood there, leaning on my physical therapists, and I cried tears of joy.

The next morning I was helped into a wheelchair and taken to the gym. There I met other patients struggling to walk again: other stroke victims and amputees, as well as people with Parkinson's disease, brain tumors and brain injuries, prosthetic hips and knees, and cerebral palsy. We formed a class, and through the weeks we urged each other on as the therapists worked with each one in turn, each one with different physical, emotional, and motivational difficulties. In our suffering and struggle, and with our progress and defeats, we bonded like a company of guerrilla warriors facing a nameless foe.

The gym is where it all happens. I still remember every patient and every therapist I ever met in the seven in-hospital and out-patient gyms I frequented. It's all there in front of you; your hopes and dreams; courage and cowardice; the genius, dedication, and compassion of the therapists; the blood, sweat, and tears of people who just days before were strangers. The gym became a state of mind, my classroom, and my home. And the therapists were my guardian angels. They were smart, skilled, tough, and forgiving. They had to kick me out at closing time.

The only thing harder than physical therapy is occupational therapy (although speech therapy may top them all, but I have no experience with it). The hands have many small muscles that perform precise movements, and it takes longer for them to respond. I was exhausted after fifteen minutes of work with my hands. Both of them were functionless; the right because of the stroke, the left because of the operations to remove dead tissue and the subsequent scarring and extensive skin grafting. At my first occupational therapy session, it was decided to concentrate on the right hand, since the left was still healing and the left shoulder and elbow needed extensive stretching before they could move even passively.

Julie, my first occupational therapist, asked me to move my right thumb. I struggled and strained. Nothing happened. She instructed me not to try so hard, but to think about moving my thumb and to try without clenching my teeth and driving my blood pressure through the roof, and not to forget to breathe. Nothing happened. We did this every day for a week, and because I was already standing by myself, however shakily, I did not regard her as insane. On Friday she wished me a good weekend and said that I should keep trying so that on Monday the thumb would move. She said that she had felt some tone in the thenar eminence at the base of the thumb. I worked all day Saturday, wondering whether my motivation was recovery or pleasing Julie. On Sunday the thumb quivered. I didn't sleep that night waiting anxiously to show my therapist.

Now, several years later, I can walk and take care of all the activities of daily living. I work productively and travel. All because of physical and occupational therapy, a field that grew up in the wake of World War II, in places like VA Hospitals and the Rusk Institute, and many others. I, like millions of others, was not left by the side of the road. Instead, we were "recalled to life."

95.

The Molecular Biology of Learning and Memory

Most of the ideas we have about the world and our civilization we have learned so that we are who we are in good measure because of what we have learned and what we remember.

—Eric R. Kandel, Nobel lecture, 2000

s you read these words, your brain is changing. But what are the changes in the brain that constitute learning and memory? Since learning and memory are complex functions, and because the brain itself is of bewildering complexity, when Eric R. Kandel decided to devote himself to this question, he adopted a reductionist approach. He wanted to keep the scientific questions he asked and the systems he worked with as simple as possible. This was a risky strategy, but it paid off big-time.

Kandel believed that by observing a simple nerve system as it learned and retained simple things he could gain insight into the mechanisms of the human brain. He chose the nervous system of the marine mollusk, *Aplysia californica*, as his experimental system. *Aplysia* is about the size of a fist. It provided the Romans with purple dye for their tunics, and it provides neuroscientists with delights and discoveries because of its remarkable nervous system. (I discovered a gastrin-like peptide in *Aplysia* in 1975.)[138] *Aplysia* has only a few neurons

376

with their cell bodies arranged in small groups, called ganglia. Each of *Aplysia*'s neurons has a name, and some of them are among the largest nerve cells known. In a great breakthrough, the neuroscience pioneers Berta and Ernst Sharrer discovered neurosecretory cells when they found the "egg-laying hormone" in *Aplysia*'s "bag cell" neuron.

Kandel studied the changes that occurred in *Aplysia*'s primitive "brain" as it learned and remembered to respond to slightly painful stimuli. In this way he was studying intricate questions with simple methods. He found that short-term retention was a function of modulation of the strength of synaptic transmission by chemicals such as serotonin, while long-term memory involved new protein synthesis. This was clearly just a beginning, the first steps on a momentous exploration.

The molecular regulation of learning and memory is now the focus of intensive study. It will have practical implications in the treatment of senile dementia, mental retardation, Alzheimer's disease, and other disorders.

Fig. 55. The infinite complexities of learning and memory are being understood through the techniques of molecular biology.

96.

Neonatology and Gerontology

At both extremes of age human physiology has unique features that require special study and care. Neonatology and gerontology are the fields that address these issues, and they have made great progress in reducing mortality and suffering.

In the Western medical tradition, attention to the care of late-term pregnancy dates back at least to the period of 715 to 673 BCE when the oldest reference to Caesarean section is found in the Roman Law of Numa Pompilius. In circa 98 to 138 CE, Soranus of Ephesus, a Greek physician practicing in Rome, wrote an early work on newborn care. The first obstetric forceps to aid delivery of the newborn was invented in England in 1650. The early eighteenth century saw the publication of average birth weight and length, and descriptions of specific diseases of newborns, like duodenal atresia and Hirschsprung's disease. Throughout the eighteenth and nineteenth centuries there were increasing descriptions of the diseases of newborns. And in 1914 the Michael Reese Hospital in Chicago opened an incubator unit for the treatment of premature infants. The very next year there was an incubator baby sideshow at the Panama-Pacific International Exposition in San Francisco.

Fig. 56. A symbolic montage representing the extremes of birth and death.

Dr. Julius H. Hess published *Premature and Congenitally Diseased Infants*, the first American textbook on prematurity in 1922. The American Academy of Pediatrics (AAP) was founded in 1930, and in 1946 Clement A. Smith published *The Physiology of the Newborn Infant*, the first American textbook of neonatology. Invasive treatment of the fetus began in 1963 with intrauterine transfusion for hemolytic disease. Neonatology as a subspecialty of pediatrics with certification examinations began in 1975. Today, infants as small as one pound are surviving, though many are so fragile that they sustain mental or physical deficits that may persist throughout their lives.

At the other end of life, gerontology is very much a product of the greatest medical advances. Increases in life span have both exposed and created the specific problems of aging. In addition to, and integrated into the care of, disease in the elderly are the financial and emotional issues that are the concerns of the gerontologist. While most physicians try to apply Ockham's razor[139] to their diagnostic approach, geriatric disease tends to be characterized by "multiplicity, duplicity, and chronicity," so that a single ("unitary") diagnosis is more often than not incomplete. As the American population ages, more research needs to be done to aid the elderly by treating physical degeneration and mental disorders and diseases such as dementia and Alzheimer's disease. Only recently has gerontology mushroomed as a field in itself.

These two fields help the weakest and most vulnerable among us. They are developing rapidly as we appreciate how human physiology changes during life. At the extremes of life's journey we are different organisms, and even the seemingly immutable signposts of birth and death are changing. Some of the great philosophical and moral issues of our time are directly related to medical advances.

97.

In Vitro Fertilization and Blastocyst Selection

What a piece of work is a man! How noble in reason, how infinite in faculty, in form and moving how express and admirable, in action how like an angel, in apprehension how like a god—the beauty of the world, the paragon of animals!

—William Shakespeare (1564–1616), *Hamlet*, 2.2, 115–17

Why did Zeus order Prometheus not to give humans fire? I suppose he felt that we would abuse its power. Knowledge can be abused, indeed it has been, but we continue to love Prometheus for his gift. The alternative would have been to live in darkness, without choice, to have been less than human. Prometheus, the cleverest Titan, who molded us from clay, knew better: the trick is in understanding our choices.

Medicine is always about knowledge and choice. Frequently, it is about life and death choices, but beginning- and end-of-life issues arouse the greatest passion. Such passion and confusion once belonged to sunrise and sunset, floods and earthquakes. Things change, and as medicine gives us more knowledge it forces more choice. To abdicate the responsibility of making choices based on this new knowledge would be tantamount to rejecting fire. In vitro fertilization and blastocyst selection are advances that give us power to intervene at the beginning of life. These techniques address fertility problems and genetic disorders.

Many couples have difficulty conceiving and/or maintaining a pregnancy. Other partners may be concerned about genetic diseases in their family tree. These problems are often due to defects within the sperm or the eggs of the prospective biologic parents, such as a high rate of aneuploidy or both parents having a gene for sickle cell anemia (sickle cell trait).

aneuploidy: Having or being a chromosome number that is not an exact multiple of the usual haploid number.

haploid number: After the primordial (precursor) germ cells (egg or sperm) populate the developing gonads of an embryo, some continue to divide by mitosis, producing more like themselves. The primordial germ cells are diploid, meaning that they have all the normal chromosomes of the organism in pairs. In humans, this means that they have twenty-two pairs of autosomes and one pair of sex chromosomes, or forty-six total. Mitosis is the name of the process whereby the cell replicates its DNA and then divides equally to result in two cells, each cell including an entire complement of DNA just like the first cell before the division (in humans, that is the forty-six total chromosomes mentioned). But if a cell is to become an ovum or sperm ready to combine with a gamete of the complementary type to produce a new organism (at first a zygote) containing the normal number of chromosomes, it must undergo a special type of cell division whereby each gamete acquires only half the diploid number. Each mature ovum or sperm must include only twenty-three single (not paired) chromosomes. Mature ova or sperm cells are haploid, indicating that the twenty-three chromosomes in their nuclei are unpaired

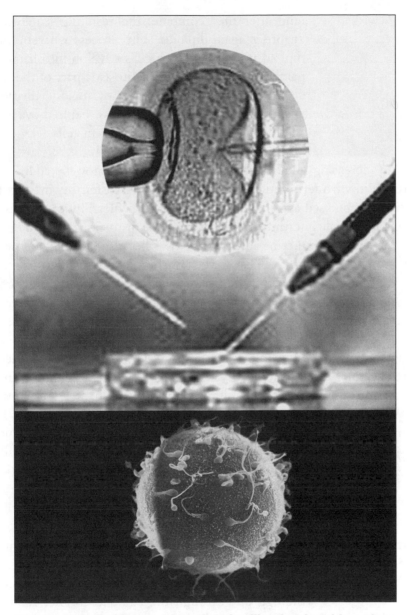

Fig. 57. The procedure of in vitro fertilization is taking place (*center*) in a petri dish with a magnified view of an ovum being penetrated by a pipette for sperm insertion (*above*). Below is an ovum in the natural process of conception in which one of numerous sperm will penetrate the cell.

(and after they combine, the resulting single-cell zygote is again diploid). The process whereby the diploid primordial germ cells develop into haploid gametes is called meiosis. Mitosis is part of the life cycle of any cell, but meiosis or meiotic division occurs only in the development of haploid ova and sperm from diploid primordial germ cells.

An increasing number of fertility problems can now be addressed through in vitro fertilization (IVF) and embryo transfer. In vitro fertilization can be used as an effective treatment for infertility of all causes except for female infertility caused by anatomic distortions of the uterus, such as severe intrauterine adhesions. This is because the egg, after fertilization in a petri dish, will have to be transferred back to develop in the uterus. In combination with the techniques of preimplantation genetic diagnosis (PGD) and blastocyst selection and transfer, an increasing number of genetic diseases can now be engineered out of a family tree.

in vitro fertilization (IVF) and embryo transfer: Involves taking eggs from the woman, fertilizing them in the laboratory with her partner's sperm and transferring the resulting embryos back to her uterus two to six (usually three or five) days later. The first IVF-baby in the world was born in July 1978 in England. Today, many thousands of children are born annually as a result of this technique. IVF was put into practice in the United States in 1981 and has resulted in 114,000 US IVF births.

intracytoplasmic sperm injection (ICSI): A variation on IVF. Instead of merely allowing sperm and eggs to come into contact in a dish, a technician physically places a sperm cell inside the egg cell through the egg membrane. ICSI is used, among other reasons, when the prospective father has some condition affecting fertility, such as a low sperm count.

preimplantation diagnosis (PGD): A test that screens for genetic flaws among genetic blastocysts used in in vitro fertilization; in other words, a method used to screen an embryo for disease. Through PGD, DNA samples from blastocysts created in vitro by the combination of a mother's egg and a father's sperm are analyzed for gene abnormalities that can cause disorders. Fertility specialists can use the results of this analysis to select only mutation-free embryos for implantation into the mother's uterus. Before PGD, couples at higher risks for conceiving a child with a particular disorder would have to initiate the pregnancy and then undergo chorionic villus sampling in the first trimester or amniocentesis in the second trimester to test the fetus for the presence of disease. If the fetus tested positive for the disorder, the couple would be faced with the dilemma of whether or not to terminate the pregnancy. With PGD, couples are much more likely to have healthy babies. Although PGD has been practiced for years, only a few specialized centers worldwide offer the procedure.

IVF is generally used in couples who have failed to conceive after at least one year of trying and who also have one or more of the following conditions:

1. Blocked genital tract passages or tubes: women with diseased or tied fallopian tubes or pelvic adhesions with distorted pelvic anatomy; men with diseased or surgically altered seminal versicles
2. Male factor infertility (low sperm count or low motility)
3. Two to six failed cycles of ovarian stimulation with intrauterine insemination
4. Advanced female age: over thirty-eight
5. Reduced ovarian reserve, that is, lower quantity (and sometimes quality) of eggs; reduced egg quantity and quality is often treated by

IVF, with or without an egg donation from another woman
6. Severe endometriosis

IVF is a rigorous technique. It demands very active participation on the part of the parents, and of course, especially the mother. Ovulation may have to be hormonally stimulated; unfertilized eggs have to be collected and examined; the uterus has to be prepared for its womb function with hormone treatments; sperm must be collected and examined; the egg must be fertilized; the fertilized egg must develop into a blastocyst while living in culture in a laboratory; blastocysts must be selected and transferred to the womb; implantation must occur. Usually more eggs are collected for fertilization than would be transferred at one time, both to increase likelihood of some successful conceptions and because the process of collecting eggs involves hormonal treatments that can be uncomfortable and risky for the woman. Any early embryos that are not transferred right away are usually stored frozen for later transfer. But many of these are not transferred. In the United States as of June 2002 there were approximately four hundred thousand embryos in storage. These unused blastocysts can be used to harvest stem cells, or they can be thrown away.

IVF requires physical and emotional strength and stamina. But the rewards are beyond compare, and the success rates are quite high. Currently, the percentage of egg retrievals resulting in live births ranges from about 60 percent down to 30 percent. The percentages fall as the age group of women being studied is older.

And now hereditary diseases, like sickle cell anemia, can be diagnosed in a blastocyst using PGD. In these (autosomal recessive) diseases we can not only insure that a child born of parents who are carriers is free of the disease but that the child will not pass on disease genes to his or her progeny.

autosomal recessive disease:	An abnormal gene on one of the autosomal chromosomes (one of the first twenty-two "non-sex" chromosomes) from each parent is required to cause the disease. People with only

one abnormal gene in the gene pair are called carriers, but since the gene is recessive, they do not exhibit the disease.

In other words, the normal gene of the pair can supply the function of the gene so that the abnormal gene is described as acting in a recessive manner. Both parents must be carriers in order for a child to have symptoms of the disease; a child who inherits the gene from one parent will be a carrier.

An increasing number of diseases including Tay-Sachs disease, Down syndrome, cystic fibrosis, thalassemia, Fanconi's anemia, Fragile X syndrome, and Huntington's chorea can be detected in the blastocyst stage. In some cases, treatment and elimination of these diseases are possible.

blastomere: A cell produced during cleavage of a fertilized egg— also called cleavage cell.

blastocyst: The modified blastula of a placental mammal, or a three-to-six-day-old fertilized ovum (egg) that consists of between three and one hundred fifty cleavage cells.

In 1999 a pair of genetically normal twins was born to a New York–area couple who are both carriers of sickle cell disease. As was reported in the *Journal of the American Medical Association*, it was the first successful use of PGD to eliminate sickle cell disease. Because sickle cell trait (one sickle cell gene) confers survival advantage to those living in a malaria region, the gene is widespread in Africa and among African Americans. One of every 625 babies born to African Americans has the blood disorder, which causes repeated episodes of severe pain, as well as a shortened life expectancy. People with the disease also have an increased susceptibility to infection, stroke, and organ failure. When both parents are carriers, they face a 25 percent chance of having a child with the disease, and half of their children may carry the gene.

This couple had aborted two early pregnancies because the disease was detected on amniocentesis. After extensive counseling and consideration, they decided to try IVF with PGD. On the second try at IVF, doctors tested seven embryos for the sickle cell gene. Four were normal, two were carriers, and diagnosis was unclear for one. Three disease-free embryos were implanted, and the woman gave birth to healthy twins after thirty-nine weeks of pregnancy.

In vitro fertilization and blastocyst selection is already one of the greatest advances because it provides children to the childless, and it's going to get better. But recent application of PGD to areas such as HLA typing and social sex selection have stoked public controversy and concern, while provoking interesting ethical debates and keeping PGD firmly in the public eye. Still, it faces the same dilemmas as all medical advances: What is its cost? Who's paying for it? And, of course, the bottom line—Is it only for rich people?

98.

Stem Cells and Regenerative Medicine

And what wine is so sparkling, what so fragrant, what so intoxicating, as possibility!

—Søren Kierkegaard (1813–1855)

Prometheus, in his phylosophical generosity, often angered Zeus. After he snuck fire in a reed to Man, the furious Zeus ordered Prometheus to be chained to a rock in the Caucasus Mountains where every day a great Eagle would come to Prometheus and eat his liver, leaving only at nightfall when the liver would begin to grow back once more, only for the process to be repeated the next day.

So began stem cell research in Western literature. We know that this experiment works, and it is from adult stem cells that the living donor grows back his liver (see advance 53). The ability to regenerate new organs, or parts of organs, when they have been damaged by disease or injury is now far beyond a mythic dream. Adult stem cells have made the medical marvel of bone marrow transplantation come true (see advance 55). And the power of our skin to heal itself, which is also dependent on adult stem cells, is a miracle which is so familiar that it is taken for granted. But the limitless potential for what is now known as regenerative medicine is in trouble in the United States.

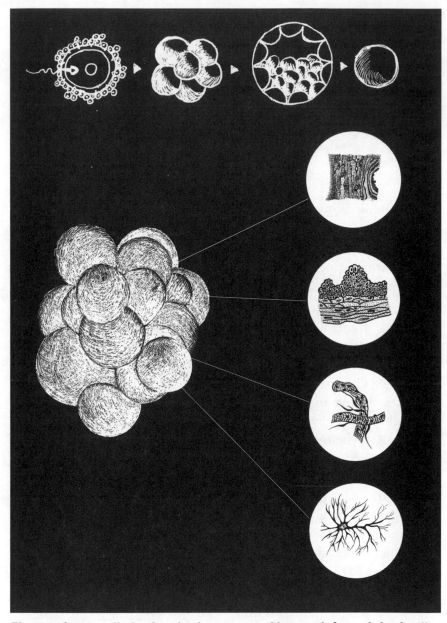

Fig. 58. Stem cells in developing ovum, *Above—left to right*, fertilized ovum, six-cell stage, blastocyst with inner cell mass, and one stem cell removed. *Below*—inner stem cell mass and its transformation into many of the different cells of the species.

In 1961 the cell type responsible for the success of bone marrow transplantation was identified as the hematopoietic stem cell. The ability of hematopoietic stem cells (HSCs) to reproduce continuously in the bone marrow and to differentiate into all the cell types found in blood makes them the outstanding example of adult stem cell research. More recently, it has been shown that HSCs can be made to produce liver cells.

In 2001 researchers reported that adult mouse bone marrow stem cells injected into mouse hearts morphed into heart muscle cells. The idea was this: by putting these cells where they were needed, they would produce the right cells. The investigators were enthusiastic about their results. Although subsequent work has failed to reproduce these results, about ten clinical trials were begun in people suffering with severe heart failure after heart attacks. In both the mouse and the human studies, crude bone marrow preparations, containing at least three types of adult bone marrow stem cells, were injected into the hearts (HSCs, which make blood cells; angioblasts, which make blood vessels; stromal cells, which make cartilage, muscle, and bone). The human trials, while not clearly showing growth of new heart muscle cells, reported about a 10 percent improvement in cardiac function. Thus, as of early 2006, it is not clear whether this approach works.

Cells that maintain the flexibility to divide and differentiate into more specialized cells of different tissue types are rare in adults. But the unlimited potential of the undifferentiated cells of the early embryo has made embryonic stem cells a primary scientific interest, as well as the best hope for curing a variety of diseases and injuries. In 1998 James Thomson of the University of Wisconsin–Madison developed the first human embryonic stem cell (ESC) cultures, and this jump-started ESC research.

ESCs are found in the earliest stage of embryonic development, when the fertilized egg has divided into between thirty and one hundred fifty cells and has not yet attached to the wall of the womb. This stage is called the blastocyst, and it exists from the fourth to the seventh day after fertilization. ESCs removed from the blastocyst can be cultured in the laboratory and can be kept proliferating indefinitely. These undifferentiated cells retain the potential to differentiate into any and all cell types of the species. But a blastocyst from which ESCs are taken is destroyed in the process, and that is the source of the controversy surrounding stem cell research.

Stem cell research and regenerative medicine are just beginning, and the future appears vast and bright. Adult and embryonic stem cells have crucial roles to play. Currently, all stem cell therapies rely on adult stem cells from bone marrow, skin, and liver. In the future, adult stem cells from other tissues may prove useful. But there are difficulties in purifying and culturing adult stem cells, and while the full range of their capacity for differentiation is not known, it is certainly more restricted than that of ESCs. ESCs are more pluripotent, and more abundant, and have an even greater contribution to make to biological understanding and human health.

Nonetheless, in 2001 President George W. Bush issued an executive order limiting federal funding for ESC research only to existing cell lines, saying, "Embryonic stem cell research is at the leading edge of a series of moral hazards."[140] Most of the cell lines cultured before the cutoff have now been contaminated. This essentially forced US stem cell scientists into private corporate laboratories, or out of the country, or into the situation of seeking funding from state sources. The avowed reason for the executive order was that harvesting ESCs destroys a human life. It is clearly related to the issues surrounding contraception and abortion, which have also been denied federal funds. The question as to whether the blastocyst is indeed a human life is surely a valid one. It deserves thought and discussion by a broad cross section of people who understand the facts. It should be understood that blastocysts are produced in attempts to produce pregnancy through in vitro fertilization, and the ones which are used for removing stem cells would otherwise be discarded because many eggs must be fertilized in IVF. But this is a question that has received little serious attention, and we are no closer to an informed consensus than we were in 2001. The implications are great. For example, if these few undifferentiated cells are a human life, why is the blastocyst more protected than the fully developed human lives that the state destroys in executions, collateral damage, economic sanctions, and neglect?

It is clear from our survey of medical marvels that decades, centuries, even millennia can pass without objective debate about critical issues. That is what happened in the sixteen hundred years between Galen and Vesalius, in the centuries when women were excluded from medical practice, in the many decades before the introduction of Medicaid and Medicare. Even now we have no real national dialogue about ESCs.

But the effort to use ESCs for the benefit of humankind goes on in this country. On April 27, 2005, the National Academy of Sciences, citing a lack of leadership by the government and the lack of debate, issued ethical guidelines for their use. They also recommended setting up a system of local and national committees for reviewing stem cell research.[141] These committees would include people from all walks of life. While these recommendations are not binding and they do not reverse the president's order, it is hoped that if they are widely read and adopted by many leading institutions, that will initiate the necessary dialogue and influence policy.[142]

99.

Prospective, Randomized, Double-Blind, Controlled Clinical Trials

Put your trust in God, my boys, and keep your powder dry!

—Colonel Blacker, *Oliver's Advice* (1834)

To know is one thing, merely to believe one knows is another. To know is science, but merely to believe one knows is ignorance.

—Hippocrates (c. 400 BCE)

For most of human cultural development, healing has been conducted through what we call ancient healing traditions. Whether Chinese, Egyptian, Indian, African, American Indian, or Australian Aboriginal medicine, these traditions have several things in common, and they have each contributed to our Western medical practice.

At some point and in some places, however, physicians and others began to rely on the repetition of empirical findings and the value of a more mechanistic process of thinking about disease and therapeutic interventions. That is not to say that traditional medicine was, or should have been, abandoned.

According to the World Health Organization, as many as 80 per-

Fig. 59. Various traditional herbs used for healing. *Left bottom to right*: opium poppy, Rauwolfia, castor-oil, meadow saffron, Madagascar perwinkle.

cent of the world's people rely on traditional medicine for their primary healthcare. And the use of traditional medicine is increasing, both in developing countries and in the "developed" world. The reasons for this are complex. Costs, both in the United States and abroad, are an important factor. But, in the United States, there is also growing dissatisfaction with the availability of Western medical care. So in spite of the great advances in Western medicine, and the greater advances that should be forthcoming, many people are looking elsewhere. The administrative complexity, impersonal atmosphere, and restricted access associated with today's healthcare have resulted in some people turning to alternatives.

We should never forget that at the heart of the healing arts and sciences there must be trust, personal commitment, and compassion. No scientific advance can replace these human values. At the same time, scientific methods are the basis of the development and evaluation of medical progress. Both of these essential elements are part of our Western tradition.

China has about 250,000 doctors trained in traditional Chinese medicine, and India has some 460,000 traditional practitioners. Treatment in these systems, and all other ancient traditions, are based on plant-derived drugs. Traditional Chinese medicine employs more than five thousand plant species. Indian healers use some seven thousand plant species. In China sales of traditional medicines have more than doubled in the last five years, while India's booming export trade in medicinal plants has risen almost threefold during the last decade. In 1990 Chinese doctors used 700,000 tons of plant material. In 1994 China's export trade in medicinal plants amounted to $2 billion. However, around the world these medicinal plants are predominantly wild growing, their habitats are shrinking, and the plants are becoming extinct faster than they can be studied. Also, these plants are not inspected or held to any standard and can vary greatly in quality. Still, we must be cognizant of the fact that aspirin, digitalis, penicillin, curare, quinine, and vitamin C are among the treatments known through traditional practice. But, in addition to the use of plants, many animal species used for ritual healing are being hunted to extinction.

The issue of trust is inherent in a paradox that lies at the core of

Western medicine and differentiates it from traditional medicine. Traditional medicine is based on belief. Western medicine is based on science. To be sure, there is some element of science in traditional medicine, since it is concerned with observation and repetition. But in traditional systems, belief trumps observation. In Western medicine, precise observation and repetition trumps belief. In Western medicine you ask why, you question what you think you know, you crave something new, and you rely more on "science" than trust. And the Western tradition evolves more rapidly.

The arrogance of culture can interfere with sharing and learning. Traditional systems make many unsubstantiated claims. This was also true of Western medicine until well into the twentieth century. Having learned through bitter experience how misleading random observation (anecdotal evidence) can be, Western medicine has recently evolved rigorous experimental methods to evaluate its findings. These methods are designed to eliminate bias ("the constant forces that make for muddlement"). Simply described, they involve strict criteria for the inclusion of experimental subjects (so that we can be as sure as possible as to what it is that we are studying), randomizing our treatment options for each subject within the experimental protocol, letting subjects and doctors who are participating in the study understand that they will not be told which treatment they are giving or receiving (double-blind testing), and including "control" treatment groups that receive inactive ("placebo") or other treatments under study. Included in this general methodology are specific procedures for randomization, safeguards for suspension of the study if one treatment arm proves to be harmful or clearly superior (as evaluated by independent study monitors), and application of appropriate statistical methods. It should be apparent that to meet these requirements the clinical study (for example, one designed to compare the efficacy of treatments) should be prospective; it must go forward from the start date, rather than look retrospectively at existing patient records. Clinical experimental design has become a dicipline of its own.

Western experimental design for clinical research is one of our greatest advances. It can be applied to virtually all treatments from any culture. But it is not completely foolproof, and it in no way reduces the requirement for compassion, integrity, and trust. The recent revelations that resulted in pulling some Cox-2 inhibitors from the market in 2005 illustrate that. People should expect unbiased scrutiny through double-blind testing of all treatments, including "natural" and traditional remedies, before agreeing to use them.

100.

Patient Advocacy

Whether 'tis nobler in the mind to suffer
The slings and arrows of outrageous fortune,
Or to take arms against a sea of troubles,
And by opposing end them?

—William Shakespeare, *Hamlet*, 3.1, 57–60

The increasing complexity of diagnosis and treatment, and especially the interposition of the insurance industry between providers and consumers of healthcare, has created the need for patient advocacy. This was unknown even a decade ago, but now almost every hospital in the United States has a patient advocate. Patient advocates, like the Patient's Bill of Rights, are a necessary, positive, and insufficient response to the adversarial atmosphere created in healthcare by capitation and other efforts to diminish staffing and service expenses in order to maximize profit.

The problem with most patient advocacy offices and organizations is that they are not independent. They may be paid by and represent the interests of the hospital from which you are seeking relief. They may represent various commercial healthcare industry representatives. They may just represent another level of expense between you and the care you need.

The point is this: Under the current circumstances, your healthcare is situated in a minefield of conflicting interests, all of which want your money. Some want more than their share, and these hide behind the fundamentalist grail of "market forces," which means the patient's well-being is second to profit. These interests now hold all the power. Patient advocates, even those who are well meaning, are not an adequate response to this crisis. Use them when you must; your family and your doctor should be your first line of support. But you had better recognize that the whole system needs fundamental change. No patient advocate can help you, or your children, or your parents with these problems until the system undergoes drastic change, until the needs of the patient become paramount once again in the healing profession.

As both a doctor and a patient, I can say that loved ones are always essential to recovery. My wife has been in attendance and advocating in my behalf every day for the past five years, both in and out of the hospital. But assuring that you get the correct medications, and that you are not left on a gurney in the hall, or that you are not discharged prematurely, should not be the job of your loved ones. The Patient Advocate is an important advance, but most of the issues that require advocacy arise from the conflicts of interest imposed by the need to create profit from your suffering.

Epilogue

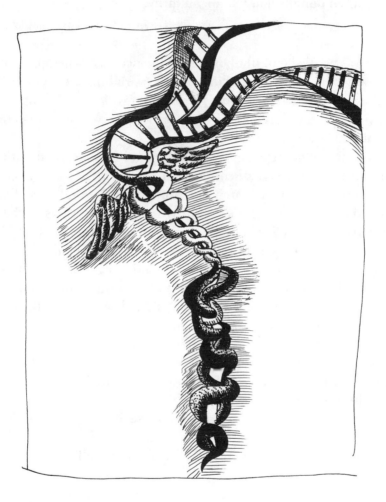

Wherever the art of medicine is loved, there also is love of humanity.

—Hippocrates (c. 400 BCE)

We have to ask ourselves whether medicine is to remain a humanitarian and respected profession or a new but depersonalized science in the service of prolonging life rather than diminishing human suffering.

—Elisabeth Kübler-Ross, Swiss-born US psychiatrist (1969)

There was no medicine until there was the need, and the opportunity, to preserve the functioning of individuals (see advance 1). As we flourished as a species, the impetus for healing grew more complex but nearly always stayed true to the highest ideals of our emerging ethical and moral belief systems.

The discipline of cognitive neuroscience is concerned with the evolutionary and structural considerations that make a brain become a mind, and scientific research now understands moral reasoning in terms of the brain's structure.[143] But whatever the reasons or the scientific basis, our feelings of empathy and our ideas of morality have brought us a long way with respect to medical advancement.

Those feelings and ideas inspired the Hippocratic Oath, Florence Nightingale, Louis Pasteur, Harold Varmus, Rita Levi-Montalcini, Gertrude Elion, Solomon Berson, Peter Medawar, David Baltimore, and most everyone who has contributed to medical practice and advances. They bind healers together throughout the world; they are the reason that my young friend (see preface) did not quit medical school, and they are why, with all our healthcare problems, healers everywhere are generally held in high esteem.

Some people feel the stirring of these moral concerns before beginning their studies, some study hoping to be inspired, most are profoundly moved through their work. They all chose medical research and/or practice.

We have tried to convey the brilliance, the utility, and the humanity of the great medical advances. We have also expressed our concerns about the present and future of medicine. Changing relationships between people in the healing arts and sciences are altering the fundamental nature of the endeavor. As Charles Dickens observes in his novel *A Tale of Two Cities*, "It is the best of times, it is the worst of times." Let us work to ensure that our better nature prevails!

Notes

1. Gustave Flaubert, *Correspondence*, vol. 1, letter to Elisa Schlesinger, January 14, 1857.

2. John Camp, *Magic, Myth and Medicine* (New York: Taplinger, 1973).

3 John Noble Wilford, "A Paleo Puzzle: Chomping without Chompers," *New York Times*, April 7, 2005, p. A11.

4. William Carlos Williams, "Old Doc Rivers," in *The Doctor Stories*, ed. Robert Coles (New York: New Directions, 1984); Anton Chekhov, "An Attack of Nerves" and "The Doctor," in *Chekhov's Doctors*, ed. Jack Coulehan (Kent, OH: Kent State University Press, 2003).

5. Antonie Leeuwenhoek, *The Collected Letters of Antoni van Leeuwenhoek*, part 15, 1st ed., ed. L. C. Palm (Netherlands: Swets & Zeitlinger, 1999).

6. "Are Doctors Losing Touch with Hands-on Medicine?" *Science Times*, Tuesday, July 13, 1999, p. F1.

7. Paul Klemperer, "Notes on Carl von Rokitansky's *Autobiography and Inaugural Address*," ed. E. Lesky, *Bulletin of the History of Medicine* 35 (1961): 374–80.

8. S. Qatayyah, *The Arabic Physician* (in Arabic), 1st ed. (Beirut: Arabic Corporation for Studies and Publication, 1984), pp. 37–43.

Some scholars insist that Ibn Nafis be recognized as the main forerunner of Servetus, Vesalius, Colombo, and Harvey in the description of pulmonary circulation as we know it today.

9. William Harvey, *On the Motion of the Heart and Blood in Animals*, trans. Robert Willis, Great Minds Series (Amherst, NY: Prometheus Books, 1993).

10. William Harvey, *An Anatomical Disputation concerning the Movement*

of the Heart and Blood in Living Creatures, trans. G. Whitteridge (Oxford: Blackwell Scientific, 1976), p. 43.

11. Herbert S. Klickstein, *Wilhelm Conrad Röntgen on a New Kind of Rays: A Bibliographical Study* (Philadelphia: Mallinckrodt Classics of Radiology, 1966); Arnulf K. Esterer, *Discoverer of X-ray: Wilhelm Conrad Röntgen* (New York: Messner, 1968).

12. William Beaumont, *Experiments and Observations on the Gastric Juice and the Physiology of Digestion*, facsimile of the original edition of 1833, together with a biographical essay, "A Pioneer American Physiologist" (Cambridge, MA: Harvard University Press, 1939).

13. Ivan Pavlov, Nobel lecture, December 12, 1904, *Nobel Lectures: Physiology or Medicine 1901–1921* (Amsterdam: Elsevier, 1967).

14. Ibid.

15. Ivan Pavlov, *Lectures on the Work of the Cerebral Hemisphere, Lecture One (1924)* (New York: Philosophical Library, 1968).

16. Pavlov, Nobel lecture, December 12, 1904.

17. Claude Bernard, *Introduction to the Study of Experimental Medicine* (New York: Dover, 1957), p. 2.

18. Charles Arthur Lovatt Evans, *Reminiscences of Bayliss and Starling* (Cambridge: Cambridge University Press, 1964).

19. Louis Pasteur, "De l'attenuation du virus du cholera des poules. C. R.," *Acad. Sci. Paris* 91 (1880): 673–80.

20. Louis Pasteur and C.-E. Chamberland, "Sur la vaccination charbonneuse. C. R.," *Acad. Sci. Paris* 92 (1881): 1378–83.

21. J. Baron, *Life of Edward Jenner* (London: H. Colburn, 1827 [vol. 1], 1838 [vol. 2]).

22. Stanley A. Plotkin, "Vaccines: Past, Present and Future," *Nature Medicine* 11 (2005): S5–11.

23. Charles Dickens, *Martin Chuzzlewit* (London: Quiet Vision, 1968).

24. Florence Nightingale, *Selected Writings of Florence Nightingale*, ed. Lucy Ridgely Seymer (New York: Macmillian, 1954).

25. Suzanne Gordon, *Nursing against the Odds: How Health Care Cost Cutting, Media Stereotypes, and Medical Hubris Undermine Nurses and Patient Care* (Ithaca, NY: Cornell University Press, 2005).

26. American Association for the History of Nursing, PO Box 175, Lanoka Harbor, NJ 08734. Ph: (609) 693-7250.

27. Peter Vinten-Johansen et al., *Cholera, Chloroform, and the Science of Medicine: A Life of John Snow* (Oxford: Oxford University Press, 2003).

28. During World War II, the IG Farben chemical company manufac-

tured the deadly Zyklon B gas that was used to kill millions of victims in Nazi gas chambers. Although IG Farben founded much of the present German chemical sector, the company itself is now being phased out. It will establish a separate foundation of several million marks to compensate slave laborers. Liquidator Volker Pollehn commented, "This is a new beginning for IG Farben."

For more information on Nazi slave labor, see the following:

—Reuters News Agency, "Fund Will Compensate Nazis' Slave Labourers," February 16, 1999.

—Israel Wire, "Lawyers Say Austrian Firms Used WW-II Slave Labor," September 17, 1998, http://www.israelwire.com/New/980917/9809177 .html.

—Howard Hobbs, "Ford Motor Co. Charged in Nazi Secret Profits on Slave Labor," *Daily Republican*, March 18, 1998, http://csufresno.com /issues_ford_slave_labor.html.

—Russ Laver, "Money and Morality: Volkswagen Has Always Insisted that It Was Following Orders from the Nazis to Use Slave Labor," *Maclean's*, September 14, 1998, http://www.quicken.ca/eng/investment/laver /morality/.

—http://www.pbs.org/wgbh/pages/frontline/shows/nazis/etc/seek.html.

—Associated Press, "'Slave' Ads Target WWII Firms: Victims of Nazis Step up Fight for Compensation," *Toronto Star*, October 6, 1999, p. A15.

29. This CDC Web site has information about a possible smallpox terrorist attack, http://www.bt.cdc.gov/agent/smallpox/index.asp.

30. Hugo Iltis, *Life of Mendel* (London: Allen & Unwin, 1932); John Gribbin, *In Search of the Double Helix* (New York: McGraw-Hill, 1985).

31. Gribbin, *In Search of the Double Helix.*

32. Jodi Wilgoren, "In Kansas, Darwinism Goes on Trial Once More," *New York Times*, May 6, 2005, p. A18.

33. See http://www.rockefeller.edu/discovery/dna/index.php.

34. National Library of Medicine, Profiles in Science, "The Oswald T. Avery Collection," p. 1, http://profiles.nlm.nih.gov/CC/.

35. http://osulibrary.orst.edu/specialcollections/coll/pauling/dna/; Gail Thompson and R. Andrew Viruleg, *Linus Pauling: A Biography* (Princeton, NJ: Woodrow Wilson National Fellowship Foundation).

36. Charles Tanford and Jacqueline Reynolds, *Nature's Robots: A History of Proteins* (New York: Oxford University Press, 2004), p. 276

37. M. Wilkins, Nobel lecture, December 11, 1961, "The Molecular Configuration of Nucleic Acids," in *Nobel Lectures: Physiology or Medicine 1942–1962* (Amsterdam: Elsevier, 1967).

38. J. D. Watson, *The Double Helix* (New York: Atheneum, 1968).

39. For a detailed account of the relationship between Alfred Nobel and Bertha von Suttner and a discussion of the Peace Prize itself, including Baroness von Suttner's reactions and opinions concerning it, see Irwin Abrams's article "Bertha von Suttner and the Nobel Peace Prize," in Bertha von Suttner, *Memoirs of Bertha von Suttner: The Records of an Eventful Life*, uthorized translation in 2 vols. (Boston: Ginn, 1910), 1:294.

40. http://online.sfsu.edu/~rone/GEessays/gedanger.htm and http://www.ornl.gov/sci/techresources/Human_Genome/elsi/elsi.shtml.

41. Francis Crick, Nobel lecture, December 11, 1962, "On the Genetic Code," *Nobel Lectures: Physiology or Medicine 1942–1962* (Amsterdam: Elsevier, 1967).

42. M. W. Nirenberg and J. H. Matthaei, *Proc. Natl. Acad. Sci.* U.S. 47 (1961): 1588; M. W. Nirenberg, J. H. Matthaei, and O. W. Jones, *Proc. Natl. Acad. Sci.* U.S. 48 (1962): 104.

43. Linda Pannozzo, "Cashing the Hunam Genome Project," *High-Grader* (May/June 2000): 10.

44. "Safeguards for Research Using Large Scale DNA Collections," *British Medical Journal*, November 4, 2000.

45. On receiving the second report of the Wellcome Research Laboratories in Khartoum, the *Daily Mail* wrote (September 25, 1906): "All Africa is going to be made perfectly habitable to the white man. Its agricultural, industrial and commercial resources will become available."

46. http://www.malariasite.com.

47. http://micro.magnet.fsu.edu/primer/techniques/index.html.

48. Sherwin B. Nuland and Richard Horton (reply), review of *The Fool of Pest: An Exchange* in the *New York Review of Books* 51, no. 5, March 25, 2004.

49. "Moralists at the Pharmacy," editorial, *New York Times*, April 3, 2005, p. WK.

50. "Actual Causes of Death in the United States: 2000," *Journal of the American Medical Association* 291, no. 10 (March 10, 2004): 1238, 1241.

51. "Give FDA Authority over Tobacco," *Chicago Sun-Times*, August 2, 2004, p. 38.

52. http://www.pfaw.org/pfaw/general/default.aspx?oid=9261#6; http://world-information.org/wio/infostructure/100437611704 /100438658297 ?opmode=contents.

53. Robert A. Levy and Rosalind B. Marimont, *Lies, Damned Lies, and 400,000 Smoking-Related Deaths*, p. 5; http://www.cato.org/pubs/regulation /regv21n4/lies.pdf.

54. "Philip Morris Report Highlights Financial Benefits to Government from Smokers' Deaths," Associated Press, July 17, 2001; "Philip Morris' Practices in the Czech Republic,"*ABC News—World News Tonight*, July 16, 2001.

55. CNN, "California Jury Awards Dying Ex-Smoker $20 Million," March 27, 2000.

56. "Tobacco Shortens Life," *In fact* 11, no. 5 (January 13, 1941): 1.

57. Ibid., pp. 3–4.

58. Laurel Druley, *Mother Jones*, "Tobacco's Belated Confessions," February 17, 1998.

59. Clive Bates and Andy Rowell, "Tobacco Control: WHO Tobacco Control Papers," University of California—San Francisco, 2004.

60. Richard Hurt and Channing Robertson, "Prying Open the Door to the Tobacco Industry's Secrets about Nicotine" *Journal of the American Medical Association* 280 (1998): 1173–81.

61. Ibid.

62. D. Wilson, *In Search of Penicillin* (New York: Knopf, 1976); Eric Lax, *The Mold in Dr. Florey's Coat: The Story of the Penicillin Miracle* (New York: Owl Books, 2005).

63. Marcia Angell, *The Truth about the Drug Companies: How They Deceive Us and What to Do about It* (New York: Random House, 2004); "At FDA, Strong Drug Ties and Less Monitoring," *New York Times*, December 6, 2004, p. 1.

64. J. F. Kihlstrom, et al., "Anesthesia, Effects on Cognitive Functions," in *Encyclopedia of Neuroscience*, 2nd ed., ed. G. Adelman (Amsterdam: Elsevier Science Publishers, 1998),

65. Douglas Heuck, "Salk Regrets Publicity That Came with His Discovery; Doctor Known for Polio Vaccine Revered by Public, Spurned by Some Scientists," *Dallas Morning News*, Bulldog ed., January 30, 1995, p. 1A.

66. Harold M. Schmeck Jr., "Jonas Salk, Whose Vaccine Turned Tide on Polio, Dies at 80," *New York Times*, June 24, 1995, p. 1.

67. Heuck, "Salk Regrets Publicity That Came with His Discovery; Doctor Known for Polio Vaccine Revered by Public, Spurned by Some Scientists."

68. Harold M. Schmeck Jr., "Albert Sabin, Polio Researcher, 86, Dies," *New York Times*, March 4, 1993, p. B8.

69. Heuck, "Salk Regrets Publicity That Came with His Discovery; Doctor Known for Polio Vaccine Revered by Public, Spurned by Some Scientists."

70. "Twenty-first Bethesda Conference: Ethics in Cardiovascular Med-

icine, October 5–6, 1989, Bethesda, Maryland," *J Am Coll Cardio* 16 (1990): 1–36.

71. Rejection letter reproduced in R. S. Yalow, *Radioimmunoassay: A Probe for Fine Structure of Biologic Systems* (Stockholm: Nobel Foundation, 1977), p. 245.

72. Personal conversation recorded in 1994.

73. This chapter is adapted from Eugene Straus, MD, *Rosalyn Yalow, Nobel Laureate: Her Life and Work in Medicine* (New York and London: Plenum, 1998), pp. 113–55.

74. Andrew Pollack, "Debate on Human Cloning Turns to Patents," *New York Times*, May 17, 2002.

75. "U.S. Germ-Research Policy Is Protested by 758 Scientists," *New York Times*, March 1, 2005, pp. 11–16.

76. Kenneth J. Carpenter, *The History of Scurvy and Vitamin C* (Cambridge: Cambrige University Press, 1986).

77. From the personal journal of Sir Richard Hawkins.

78. Louis H. Roddis, *A Short History of Nautical Medicine* (New York: Hoeber, 1941), p. 53.

79. Henry James, preface to *What Maisie Knew* (London: Penguin Books, 1986).

80. Samuel S. Fitch, *A System of Dental Surgery* (Philadelphia: Carey, Lea & Blanchard, 1835).

81. P. B. Medawar, *Cellular Aspects of Immunity* (Boston: Ciba Foundation Symposium, 1960), pp. 134–49.

82. UNOS Web site, http://www.unos.org/.

83. Jeffrey P. Kahn, "Studying Organ Sales: Short Term Profits, Long Term Suffering," a study published by the *Journal of the American Medical Association* on October 2, 2002, which looks at the effects of kidney sales done in a large city in southern India, where, according to the article, the practice has been going on for more than a decade.

"Online Shoppers Bid Millions for Human Kidney," CNN.com, September 3, 1999—"Online shoppers who frequent the Internet auction site eBay are used to seeing a wide variety of things for sale. On a given day, anything from a grandfather clock to 40 acres of Wyoming ranch land can be found among the thousands of items up for bidding. Still, the notice of a 'fully functional kidney' put up on the site last week created a stir. It also brought in bids of more than $5.7 million before the company intervened to block the sale."

84. The statistics in this chapter are based on data available from the Organ Procurement and Transplantation Network as of October 31, 2004.

85. Personal communication with the patient, Fran Epstein.

86. M. J. Laughlin et al., "Outcomes after Transplantation of Cord Blood or Bone Marrow from Unrelated Donors in Adults with Leukemia," *New England Journal of Medicine* 352, no. 22 (November 25, 2004): 2265–75.

87. "Pregnancy, No Insurance Don't Mix, Midwives Say during Cover the Uninsured Week," *U.S. Newswire*, May 4, 2005.

88. Division of Reproductive Health, National Center for Chronic Disease Prevention and Health Promotion, CDC, "Summary of Noifiable Diseases," *MMWR* 48, no. 38 (October 1, 1999): 849–57.

89. J. M. McGinnis and W. H. Foege, "Actual Causes of Death in the United States," *Journal of the American Medical Association* 270 (1993): 2207–12.

90. "Study Says Hospitals Could Lose Millions under Pataki Budget," *Plattsburgh Press-Republican*.

91. "An Intensive Effort to Save Lives; Lutheran Hospital's Pilot ICU Program Succeeds in Reducing Death Rates," *Rocky Mountain News (Denver)*, November 15, 2003.

92. http://www.sccm.org/professional_resources/administrator_toolbox/Documents/BestPracticeModel.pdf.

93. http://encyclopedia.laborlawtalk.com/Intensive_care_unit.

94. A. Tomasz "Multiple-Antibiotic-Resistant Pathogenic Bacteria: A Report on the Rockefeller University Workshop," *New England Journal of Medicine* 330, no. 17 (April 28, 1994): 1247–51.

95. CDC Division of Bacterial and Mycotic Diseases: Antibiotic Resistance, www.cdc.gov/drugresistance/community.

96. Rita Levi-Montalcini, "Autobiography," *The Nobel Prizes*, ed. Wilhelm Odelberg (Stockholm: Nobel Foundation, 1987).

97. http://www.westviewhospital.org/renal_carc_dialysis.htm.

98. Dori Schatell, "Home Dialysis, Home Dialysis Central, and What You Can Do Today," *Nephrology Nursing Journal* (March 1, 2005).

99. Gerard Manley Hopkins (1844–1889), "The Windhover," in *Poems* (1918).

100. "In Genome Race, Government Vows to Move Up Finish," *New York Times*, September 15, 1998, p. F3.

101. "Human Genome Project: Completes Human Genome Map," *American Health Line*, April 15, 2003.

102. Andrew Pollack, "Celera to Quit Selling Genome Information," *New York Times*, April 27, 2005, business section.

103. Ibid.

104. R. Luft et al., "A Case of Severe Hypermetabolism of Nonthyroid

Origin with a Defect in the Maintenance of Mitochondrial Respiratory Control: A Correlated Clinical, Biochemical, and Morphological Study," *Journal of Clinical Investagation* 41 (1962): 1776–1804.

105. Sigmund Freud, "Lecture XXXV: A Philosophy of Life," in *New Introductory Lectures on Psycho-analysis* (London: Hogarth Press, 1933).

106. Sigmund Freud, *The Standard Edition of the Complete Works of Sigmund Freud*, 24 vols., ed. James Strachey et al. (London: Hogarth Press and the Institute of Psychoanalysis, 1953–1974).

107. Freud, "Lecture XV."

108. Freud, "Lecture XVI."

109. S. Scheidlinger, "The Group Psychotherapy Movement at the Millennium: Some Historical Perspectives," *International Journal of Group Psychotherapy* 50, no. 3 (July 200): 315–39.

110. *Principles of Drug Addiction Treatment: A Research-based Guide*, National Institute on Drug Abuse (NIDA), National Institutes of Health (NIH), NIH Publication No. 99-4180, 1999, p. 5.

111. "Religion in America: Moral Re-armament Taking More Public View," UPI, July 8, 1988.

112. Rational Recovery, http://www.rational.org/.

113. http://www.whitehousedrugpolicy.gov/publications/factsht /methadone.

114. Claude Bernard, *Introduction to the Study of Experimental Medicine* (New York: Dover, 1957), p. 2.

115. *Human Radiation Experiments Associated with the U.S. Department of Energy and Its Predecessors*: This document contains a listing, description, and selected references for documented human radiation experiments sponsored, supported, or performed by the US Department of Energy (DOE) or its predecessors, including the US Energy Research and Development Administration (ERDA), the US Atomic Energy Commission (AEC), the Manhattan Engineer District (MED), and the Office of Scientific Research and Development (OSRD). The list represents work completed by DOE's Office of Human Radiation Experiments (OHRE) through June 1995. The experiment list is available on the Internet via http://www.ohre.doe.gov. The home page also includes the full text of *Human Radiation Experiments: The Department of Energy Roadmap to the Story and the Records* (DOE/EH-0445), published in February 1995, to which this publication is a supplement: www.eh .doe.gov/ohre/roadmap/achre/chap5_2.html.

Copies of *Advisory Committee on Human Radiation Experiments: Final Report* (stock no. 061-000-00-848-9) and copies of its three supplemental volumes can be obtained from the superintendent of documents, US Government Printing Office, 202-512-1800. This report is also available from Oxford University Press under the title *The Human Radiation Experiments*, 1996.

See also J. M. Harkness, "Nuremberg and the Issue of Wartime Experiments on US Prisoners: The Green Committee," *Journal of the American Medical Association* 276 (1996): 1672; J. Katz, "The Nuremberg Code and the Nuremberg Trial: A Reappraisal," *Journal of the American Medical Association* 276 (1996): 1662; and V. W. Sidel, "The Social Responsibilities of Health Professionals: Lessons from Their Role in Nazi Germany," *Journal of the American Medical Association* 276 (1996): 1679.

116. http://ohsr.od.nih.gov/guidelines/nuremberg.html.

117. J. A. Barondess, "Medicine against Society: Lessons from the Third Reich," *Journal of the American Medical Association* 276 (1996): 1657.

118. L. E. Sullivan et al., "The Prevalence and Impact of Alcohol Problems in Major Depression: A Systematic Review," *American Journal of Medicine* 118 (2005): 330–41.

119. T. Wolinsky and H. Brune, *Serpent on the Staff* (New York: Putnam, 1994). In this book, *Chicago Sun-Times* reporters Wolinsky and Brune discuss and meticulously document the AMA's history, organization, relationships with drug and tobacco industries, positions on Medicare and Medicaid, AIDS, minorities and women in medicine, and other important health issues.

Jerome Kassirer, *On the Take: How Medicine's Complicity with Big Business Can Endanger Your Health* (Oxford: Oxford University Press, 2004). In this book Kassirer, the former editor in chief of the *New England Journal of Medicine*, documents, with well-referenced examples, how conflicts of interest, primarily financial in nature, have infiltrated all areas of the medical profession.

Donald L. Barlett and James B. Steele, *Critical Condition: How Health Care in America Became Big Business—And Bad Medicine* (New York: Doubleday, 2004).

120. Paul Krugman, "A Private Obsession," *New York Times*, April 29, 2005, p. A25.

121. General Accounting Office report, *Administrative Waste in U.S. Health Care System* (June 5, 1991).

122. Weiss Ratings, Inc., http://www.weissratings.com/; GAO report, *Comparison of Private Agency Ratyings for Life/Health Insurers* (September 1994).

123. "Unconventional Medicine in the United States: Prevalence,

Costs, and Patterns of Use," *New England Journal of Medicine* 328, no. 4 (1993): 246–52

124. "Healthcare Chief Gets $1 Billion," *New York Times*, June 15, 1996, p. A32.

125. The Census Bureau's International Data Base (IDB)—Table 010—1998 data on life expectancy and infant mortality rate: http://64.233 .161.104/search?q=cache:WA5C4Hzy_H0J:library.thinkquest.org/16665/ex pectancy.htm+life+expectancy+by+country&hl=en.

126. AAMC's Executive Council, "Final Report in the Status of Women in Medicine," *Academic Medicine* 77 (2002): 1043–61.

127. Committee on Graduate Medical Education (COGME), "Twelfth Report: Minorities in Medicine," May 1998; AAMC, "AAMC's Executive Council Final Report on the Status of Women in Medicine," *Academic Medicine* 77 (2002): 1043–61.

128. Jeanne Achterberg, *Woman as Healer* (Boston: Shambhala, 1991).

129. *The Malleus Maleficarum of Kramer and Sprenger*, ed. Montague Summers (New York: Dover, 1971).

The Malleus Maleficarum (The Witch Hammer), first published in 1486, is arguably one of the most infamous books ever written, due primarily to its position and regard during the Middle Ages. It served as a guidebook for Inquisitors during the Inquisition and was designed to aid them in the identification, prosecution, and dispatching of witches.

130. Anne C. Rose, *Transcendentalism as a Social Movement, 1830–1850* (New Haven, CT: Yale University Press, 1981).

131. A. Flexner, *Medical Education in the United States and Canada: A Report to the Carnegie Foundation for the Advancement of Training* (New York: Carnegie Foundation for the Advancement of Training, Bulletin No. 4-1910, 1910).

132. "Harvard President Apologizes for Remarks on Gender," *New York Times*, January 20, 2005, p. A1.

133. W. Michelfelder, *It's Cheaper to Die: Doctors, Drugs and the American Medical Association* (New York: G. Braziller, 1960); D. F. Stroman, *A Social and Ethical Analysis of Selected Practices of the American Medical Association 1945–62* (Boston: Boston University Press, 1966).

134. http://www.enviroliteracy.org/article.php/63.html.

135. State of New York Department of Health, *Investigation of Cancer Incidence and Residence Near 38 Landfills with Soil Gas Migration Conditions: New York State, 1980–1989* (Atlanta: Agency for Toxic Substances and Disease Registry, June 1998).

136. To Avril, "Ash Finds a Heap to Call Home; The 16-Year Saga of

Phila.'s Infamous Load of Garbage Ends in Pa.," *Philadelphia Inquirer*, June 28, 2002.

137. "With No Room to Dump, U.S. Faces Garbage Crisis," *New York Times*, June 29, 1987; "Trash—Burning Question in N.Y.," *Deseret News (Salt Lake City)* March 23, 2002.

138. E. Straus, H. Gainer, and R. S. Yalow, "Molluscan Gastin: Concentration and Molecular Forms," *Science* 190 (1975): 687–89.

139. Sir William of Ockham was a fourteenth-century logician who taught that the least complex answer was most probably correct. In medicine, that translates into the rule of the "unitary diagnosis," or that if you can explain many signs and symptoms with one.

140. The President's Council on Bioethics, Washington, DC, January 2004, http://www.bioethics.gov.

141. National Research Council and Institute of Medicine of the National Academies—Committee on Guidelines for Human Embryonic Stem Cell Research, *Guidelines for Human Embryonic Stem Cell Research* (Washington, DC: National Academies Press, 2005), pp. 1–178.

142. Editorial, "Stem Cell Frontiers," *Chicago Tribune*, November 7, 2005.

143. See Michael Gazzaniga, *The Ethical Brain* (New York: Dana Press, 2005), and Rita Carter, *Mapping the Mind* (Berkeley and Los Angeles: University of California Press, 1999).

List of Illustrations

—Images on pages 14, 20, 34, 38, 46, 54, 58, 79, 81, 95, 117, 118, 123, 133, 140, 149, 154, 169, 206, 219, 233, 242, 371, 383, and 395 (used in montages) adapted from photographs with permission from Photo Researchers.

—Fig. 49: From THE WATER PLANET by Lyall Watson and Photographs by Jerry Derbyshire, copyright © 1988 by Lyall Watson. Photograph of "Iceberg" in montage copyright © 1988 by Jerry Derbyshire. Used by permission of Crown Publishers, a division of Random House, Inc.

—Figs. 31 and 33 are constructed from the data of Donald Steiner.

—All historical images are in the public domain.

—All line drawings and montages are original works created by Bette Korman.

Index

419